THE DRUIDRY
Handbook

The DRUIDRY
Handbook

Spiritual Practice Rooted in the Living Earth

JOHN MICHAEL GREER

WEISERBOOKS
San Francisco, CA / Newburyport, MA

First published in 2006 by
Red Wheel/Weiser, LLC
With offices at:
500 Third Street, Suite 230
San Francisco, CA 94107
www.redwheelweiser.com

ISBN-10: 1-57863-354-0
ISBN-13: 978-1-57863-354-8

Library of Congress Cataloging-in-Publication Data
Greer, John Michael.
The Druidry handbook / John Michael Greer.
    p. cm.
Includes bibliographical references (p.   ) and index.
ISBN 1-57863-354-0 (alk. paper)
1. Druids and druidism. I. Title.
BL910.G75 2006
299'.94—dc22
                2005023545

Cover and book design by Anne Carter
Typeset in Perpetua

Printed in Canada
TCP
10 9 8 7 6 5 4 3

For Sara
Companion on the forest paths

CONTENTS

Part III:

FOREWORD

A WORK OF ALCHEMY

When you die, only three things will remain of you, since you will abandon all material things on the threshold of the Otherworld: what you have taught to others, what you have created with your hands, and how much love you have spread. So learn more and more in order to teach wise, long-lasting values. Work more and more to leave to the world things of great beauty. And Love, love, love people around you for the light of Love heals everything.

FRENCH DRUID TRIAD, FRANÇOIS BOURILLON

THE STORY OF DRUIDISM CAN BE DIVIDED INTO three phases: the ancient Druids, who left only traces of their teachings; the Revival Druids of the eighteenth and nineteenth centuries, who left behind a great deal of literature; and modern-day Druids, who from the mid-twentieth century onward have developed Druidry into one of the leading alternative spiritualities in the world today.

Modern Druidry has faced considerable difficulty establishing itself as a valid and viable spirituality. Critics have suggested that its two earlier manifestations are fatally flawed as sources. The amount of material inherited from the ancient Druids is minimal, and the material inherited from the Revival Druids is often spurious and incomprehensible. Despite these difficulties, contemporary followers of Druidry have managed to find inspiration and guidance in both these sources. In fact, they have realized that, to a great extent, a spirituality that works is one that is made rather than simply found.

Born, as we are, into a consumer culture, we may long for a ready-made spirituality—one that is fixed and "pure." Any study of the history of religion and spirituality shows, however, that even apparently revealed religions are actually made up of a number of strands of influence that continue to change to keep the spiritual tradition vital and not succumb to the arteriosclerosis of fundamentalism.

It has been fashionable, particularly in the United States, to dismiss the entire period of Revival Druidry as an aberration best forgotten. The flamboyant and witty author of *Real Magic*, Isaac Bonewits, once Archdruid of the ADF, the largest Druid group in America, refers to the three phases of Druidism as *palaeopagan*, *mesopagan*, and *neopagan*. He believes that todays' neopagan Druids must reject the mesopagan (Revival) period altogether, and instead attempt to research and restore as much of palaeopagan Druidry as possible. A similar approach is taken by Celtic Reconstructionism, a movement also originating in the United States, that likewise rejects the contributions of Revival Druidry.

Bonewits's and the reconstructionists' arguments are convincing. Much of Revival Druidry attempts to reconcile Druidry with Christianity. Moreover, much of it is based on the work of Iolo Morganwg, which, although originally presented to the world as authentic ancient teachings, was subsequently discovered to derive from manuscripts forged by Iolo himself.

We reject aspects of our past, however, at our own peril. The search for an authentic Druidry that rejects Revival Druidry altogether makes a threefold error. First, it assumes that all of Iolo's contributions were generated in his own mind, when the evidence suggests that he may well have gathered scraps of authentic lore that he wove into the fabric of his creations. Second, it condemns Iolo's contributions and the writings of all Revival Druids simply because they may have been created in recent times rather than in the ancient past, when we know that age is not a determinant of value or wisdom. Third, it rejects all Revival Druidry because some of its proponents, and most particularly Iolo, lied about the provenance of their offerings, when we know that it is a curious feature in the history of spiritual movements that personal creativity and inspiration, which can be valuable and helpful to others, is often misrepresented as deriving from ancient, and therefore more authoritative, sources. European occultists of the eighteenth, nineteenth, and twentieth centuries seemed particularly prone to this misrepresentation: the early histories of Theosophy, the Golden Dawn, Wicca, and some of the Rosicrucian movements spring to mind. Similarly, there are those who falsely claim authority by flaunting bogus doctorates.

The consequence of this threefold error is that, despite the problems presented by Revival Druidry, reconstructionism remains the bastion of a minority, while the forms of contemporary Druidism that flourish in the world today include the Revival as one of their sources of inspiration. Why is this? What is there of value in its teachings?

We must look to America for the answer. Revival Druidry was sustained in America; it was lambasted in America; and now it is being redeemed in America. The greatest collection of literature in the world on Revival Druidry is housed in the New York City Public Library. For a decade or more, Bonewits on the East Coast and the reconstructionists on the West, have dismissed Revival Druidry as an aberration. Now, from a peaceful valley in Oregon, John Michael Greer has taken on the task of redeeming it.

Alchemy is the art of revealing the gold that lies concealed within apparently valueless matter. In this book, Greer performs alchemy. He discovers and articulates the gold that lies hidden within the obscure texts of the Revival Druids. In what amounts to a master stroke, he takes what has long been seen as a core weakness and reveals it as a valid spirituality. He succeeds with consummate skill in offering a perspective that redeems Druidry's core, revealing it to be a heritage around which we can grow and build a vital and dynamic practice rooted in the living earth.

Philip Carr-Gomm
Lewes, Sussex
Lughnasadh 2004

ACKNOWLEDGMENTS

THIS BOOK WOULD NOT HAVE BEEN POSSIBLE without a great deal of help from other people in the Druid community. Special thanks must go first to Philip Carr-Gomm, Chosen Chief of the Order of Bards Ovates and Druids (OBOD), for his constant friendship and encouragement, for the Foreword to this book, and for making possible a research trip to British Druid archives in the summer of 2003. The Order's Patroness, Dwina Gibbs-Murphy; its Scribe, Stephanie Carr-Gomm; its Archivist, Gordon Cooper; and many other members of the Order in Britain helped make that research trip, and my involvement in OBOD generally, a memorable and inspiring experience. My deepest thanks to them all.

My introduction to the world of modern Druidry came by way of Clann na Mara Intire, the OBOD seed group in Seattle. Thanks are owed here to Corby Ingold, Mark O'Kelly, Louise Candage, and Bob and Roxanne Duniway. My Druid studies led me to the Druid fellowship Ar nDraiocht Fein (ADF), whose extraordinary community of scholars and practitioners taught me to expand my sense of the tradition and deepen my knowledge of its historical and spiritual dimensions. Robert Barton, Ian Corrigan, Todd Covert, Erik Dutton, Robert "Skip" Ellison, Jenni Hunt, David and Toby Kling, Jeremy Mallory, Ceisiwr Serith, and many other past and present ADF members helped expand this dimension of my Druid journey. The Whiteoak online Druid community also taught me important lessons: Jan Graham-MacMillan, Ellen Evert Hopman, Alexei Kondratiev,

Rebecca Terrill, and many other people on the Whiteoak list shared their knowledge and insights freely.

In the Ancient Order of Druids in America (AODA), Archdruid John Gilbert was instrumental in bringing me into the Order and giving me the opportunity to contribute to it, as an Archdruid of the Grand Grove and as Grand Archdruid. I also owe thanks to Rev. Betty Reeves, Past Grand Archdruid, and to all the members of the Grand Grove and the AODA for their insights and their support. Special thanks go to Cathy Aurand, Patrick and Mary Claflin, and all the members of the Tir na Rogue Druid community in Medford, Oregon, for their hospitality, encouragement, and support.

The first draft of this book was read by Philip Carr-Gomm, Gordon Cooper, and John Gilbert. Their comments helped me clarify and improve the material and its presentation in more ways than I can count. Of course, all remaining flaws are my responsibility alone.

Finally, thanks go to my fellow Druid, traveling companion, and source of inspiration, my wife Sara, who was involved in this book at every stage, from first conception to final proofs. This book is dedicated to her.

INTRODUCTION

A DRUIDRY FOR TODAY

WHAT DOES IT MEAN TO BE A DRUID TODAY? It's a valid question, and this book will try to answer it. It's also a controversial question that raises challenging issues at the intersection of spirituality, history, and myth. To the world of conventional scholarship, modern Druidry is an oxymoron, for Druids are a thing of the past—the extinct priesthood of a barbarian culture relevant only to specialists. To the mainstream religions and philosophies of the West, modern Druidry is an absurd anachronism—a cult that turns its back on progress and the modern world to embrace an archaic reverence for trees and stones.

Yet for more than three centuries, the image of the Druid has haunted the imagination of the Western world, offering a vision of humanity as children and students of the living Earth—not her adversaries and would-be conquerors. That vision has long challenged conventional ideas of progress, learning, and religion. Literature, poetry, music, art, protest, social criticism, and alternate visions of history have often raised the Druidry of old as a banner in the face of dogmatic religion and spiritually barren materialism. Yet the most powerful of these challenges are crafted by people who take the ancient Druids as a living example and follow a Druid path in the midst of the modern world.

These modern Druids draw inspiration from ancient Celtic wisdom traditions, but their traditions have not come down from the Druids of old. Instead, they accept and celebrate their heritage as part of a modern spiritual movement open to growth, change, and new insights, rooted in the green Earth and embracing the contributions of many peoples and times. Modern Druids learn from ancient teachings, the developing tradition of modern Druidry itself and the ever-changing lessons of the living Earth. They embrace an experiential spirituality and forsake rigid belief systems for disciplines of inner development and personal contact with the realms of nature and spirit.

Many people today yearn for a way of life in harmony with the green Earth, a spirituality that celebrates nature, and a vision of the future more honest and hopeful than the bleak realities of the present and the tawdry plastic Utopias promoted by advertising and political and economic parties. Sometimes this yearning leads to attempts to reject modern life in favor of an idealized past seen through the rose-colored glasses of romantic fantasy, but such efforts lead nowhere. What's needed in the crisis of the present age is a path that brings nature and spirit together in the here and now, in the world we actually inhabit. What's needed is a spirituality that offers effective tools and practical guidance for the hard, but necessary, task of transforming ourselves and our world.

Druidry offers such a path. It isn't a panacea for the world's problems. As with any spiritual tradition, its potential varies with the capacities of the very human individuals who practice and teach it. Still, as a living tradition of nature spirituality, it has much to offer the modern individual and the modern world.

This book addresses those who are dipping a tentative toe in the clear waters of the Druid tradition, those plunging into its depths headfirst, and those who fall somewhere in between. Its focus is experiential, personal, and practical. Like much of Druid wisdom, it unfolds in three parts. Part I, *A Quest for Ancient Springs: The Druid*

Tradition in the Modern World, outlines the history of the modern Druid movement and explores the context of Druidry as a contemporary spiritual path. Part II, *The Wisdom of the Stone Circle: Nine Druid Concepts*, covers some of the core symbols and teachings of Druid philosophy and provides material for the meditation exercises central to Druidry as a living path. Part III, *The Ways of the Sacred Grove: Initiation into the Druid Tradition*, presents the central themes of Druid practice—movement toward a life in harmony with the Earth, celebration of nature's cycles and powers through seasonal rituals, and the development of body and mind through meditation. It discusses the three central facets of Druid training: the Earth Path of nature awareness and natural living, the Sun Path of seasonal celebration, and the Moon Path of meditation. Those who feel themselves called to follow the Druid's path can use this material as a starting point for their own journeys into the green realms of today's Druidry.

Like every living spiritual tradition, Druidry grows and changes constantly. Indeed, different branches of the movement have developed unique approaches. Druid orthodoxy is very nearly a contradiction in terms. Ask three Druids the same question, a common joke has it, and you'll get at least six different answers. Nothing in this or any other book on Druidry should be treated as infallible or taken on faith. The only source of Druid teachings that deserves such reverence is the world of nature itself.

How to Use This Book

Those who know the flowing complexity of the Druid tradition won't be surprised to learn that this book can be read in several different ways. People come to Druidry and books on Druidry from many directions, and their paths through this book will vary accordingly.

For those who are curious about Druidry but don't yet know much about it, or who are trying to decide if the Druid tradition speaks to them, the best approach is to read through the book from front to back. The history of the modern Druid tradition covered in Part I offers a good introduction to Druidry as a living spiritual movement, since nearly all the themes of the modern tradition have deep roots in the past. The symbolism and cosomology presented in Part II gives a glimpse into the deeper philosophy of the tradition, while the practical material in Part III takes its meaning from the play of historical change and timeless wisdom framed by the first two parts.

For those who already know they want to follow the Druid path, on the other hand, the opening section of Part III is the best place to start. Though it has an extraordinary history and a wealth of lore and philosophy, Druidry isn't primarily an intellectual path. Its core is experiential and best reached through the practice of nature awareness, seasonal celebration, and meditation. Part III describes these practices. The traditional Druid methods of meditation require some familiarity with the lore presented in Part II, however, and readers with practical work in mind should study this material next. The historical perspectives in Part I have less to offer the Druid-in-training, although the history of the Druid movement has important lessons to offer Druids today.

Those who are following a Druid path as part of a Druid order may use this book in yet a different way. It contains the core elements of the Druid tradition and is designed for use by solitary students as a system of Druid training. However, it's also the main textbook for the Druid Apprentice study program of the Ancient Order of Druids in America (AODA), a traditional Druid order I head as Grand Archdruid. An Afterword offers information about AODA and its course of study for those who are interested in exploring Druidry further.

The Path of Initiation

Today, the term "initiation" has come to mean a formal ritual marking the beginning of a spiritual path. Many Druid organizations have rituals of initiation—in fact, sometimes several of them, for different grades or levels. Some orders require the efforts of a Druid grove with trained members, while others include rituals for self-initiation in their instructional coursework. The Ancient Order of Druids in America falls into both camps; it confers its three degrees either in an AODA grove or in solitary ritual form.

Since this book is meant to be useful to Druids whether or not they choose to belong to a Druid order, it outlines a different path. Too many people forget that the word "initiation" simply means "beginning." They mistakenly assume that the simple act of passing through an initiation ceremony by itself makes them Druids. Records of ancient Druid schools, where some students spent twenty years mastering the Druid curriculum, provide a useful corrective to this sort of thinking.

In a broader sense, becoming a Druid is a lifelong work. Initiating yourself as a Druid involves reshaping your relationship to the Earth, to the spiritual powers of nature, and to yourself. A year of practice with the basic skills of the Druid tradition is a good starting point. If you choose to follow the Druid way outlined in this book, you can begin your initiation by celebrating one of the Druid holy days (see chapter 8), and end it at the same holy day a year later. Before you begin, read and thoroughly review the material on each of the three paths given in Part III. Be ready to start at least one other part of your Druid training that same day and proceed from there.

During your initiatory year, keep track of your Druid experiences and practices in a journal; any convenient notebook will do. Keeping a Druid journal is a useful way to track your progress and can be a source of perspective. Bring it up to date each day if at all possible, and review it at intervals. If you decide to pursue Druid initiation and study in the Ancient Order of Druids in America, your journal will also give you the raw material for your first-degree examination.

The time and effort required to work through your initiatory year aren't excessive. The Moon Path (see chapter 9) starts out with just ten minutes of meditation two or three times a week, increasing to twenty or thirty minutes daily by the end of the year. Two or three hours a week are enough for the other paths, with an additional hour or so for each of the Druid holy days. Turning off the television for half an hour each night or setting the alarm thirty minutes earlier each morning can free up more than enough time to complete all the material in this book with room to spare.

Still, a year of Druid practice is a test as well as an initiation. After a few days or weeks of solitary practice, the glamor of "being a Druid" can start to wear thin. The Earth Path described in chapter 7 makes life somewhat less convenient; the Sun Path described in chapter 8 may involve awkward scheduling; the Moon Path described in chapter 9 can be tiresome and frustrating at times. All three demand patience and a willingness to learn from your mistakes. You can always find good excuses to skip practices, and plenty of other things can seem more important than Druid training on any given day. You must face these discomforts and excuses and make a choice between the romantic image of Druidry and the harder, but more rewarding, realities of Druidry as a living spiritual path. Put another way, the choice is between being a Druid and actually becoming one.

This is an important choice; it can't be forced and it can't be faked. If you find that your interest in Druidry stops where the hard work of Druidry begins, it's best to realize that, accept it, and do something else with your time. On the other hand, you may find that living the Druid way makes you feel as if you are finding your way back to a home you never knew you had. Many Druids share this experience.

For them, despite the occasional difficulties and frustrations of the journey, the process of learning Druidry becomes an adventure unlike any other.

PART I:

A QUEST FOR ANCIENT SPRINGS

The Druid Tradition in the Modern World

I

CIRCLES AND STANDING STONES

MOST SPIRITUAL TRADI-
TIONS TRACE THEIR HISTORY BACK
TO A REVELATION THAT GAVE THEM all
the answers. Druidry is a different kind of spirituality,
however, and its history tells a different tale. It doesn't
claim to have all the answers—in fact, it's much more
interested in asking questions—and it didn't start with a
revelation. It started with a quest.

For thousands of years the British have inhabited a
land shaped by ancient hands. Tall stones loom out of the
grass, alone or in patterns. Long barrows and round barrows mark the skyline or rise in the middle of pastures
and fields. Odd customs linger around some of these, preserved by habit or a vague sense that ill luck will follow if
they are neglected. Living close to the shapes of the land
and the echoes of tradition for countless generations,
country folk in the seventeenth century came barely to
see them at all.

When the gentleman scholar John Aubrey rode up to
the little Wiltshire village of Avebury on a cold January

day at the beginning of 1649, he had no idea that he was about to enter the greatest surviving temple of prehistoric Europe. Locals knew that hundreds of massive stones lay scattered across Avebury's fields, but the broader pattern they formed—a vast triple circle inside a bank and ditch of gigantic scale—lay hidden in plain sight by its sheer size. Familiar with Stonehenge, a day's hard ride to the south, and fascinated with the lore and legends of the English countryside, Aubrey saw meaning where others noticed only large inconvenient stones. He later commented that Avebury "did as much excell Stoneheng, as a Cathedral does a Parish church."[1]

Aubrey was a new kind of scholar, poised between the fading Renaissance and the first stirrings of the modern world. Like the great minds of the Renaissance, he sought what we now call a holistic view of things. Like the first proponents of the Scientific Revolution, many of whom were among his friends, he drew information first and foremost from the world around him, breaking free of the obsession with written texts that had shackled the Middle Ages. The book he valued most, to use a revealing metaphor of the time, was the Book of Nature.

Like many of his contemporaries, Aubrey stood between a fading age and a dawning one in another way. He rode into Avebury less than a year after the end of the Thirty Years War, a nightmare struggle between Catholics and Protestants that left most of Central Europe in smoking ruins. A few weeks after his visit, England's King Charles I was beheaded in a savage finale to the English Civil War, a political and religious struggle that ripped apart the core of English society. These explosions capped a century of ferocious conflict over religion that left the moral claims of organized Christianity in tatters.

Many people of good will, horrified by the carnage, turned to the scientific materialism offered by philosophers such as René Descartes. Yet the "mechanical philosophy," as it was called, had problems of its own. Founded on a materialism that was just as dogmatic and rigid as the religions it opposed, the new vision of a clockwork universe set in motion by an absentee god threatened to empty the world of meaning and make humanity lose touch with its own spiritual possibilities. Reduce the cosmos to lifeless atoms colliding in a void, insightful people had already realized, and every human and spiritual value gives way to a universe ruled by blind necessity and brute force.

The forced choice between murderously dogmatic religion and spiritually barren materialism drove many people to look for a third option. Whispers of other possibilities were in the air, for those with ears to hear. Ancient manuscripts rediscovered by Renaissance scholars offered glimpses of long-forgotten spiritual paths. Travelers from distant countries brought back tales of what we now call shamanic traditions in North America, Lapland, and the eastern reaches of the expanding Russian Empire.

Another potent factor lay closer to home. Monuments such as Avebury and Stonehenge posed a silent challenge to the British culture of Aubrey's time. As Aubrey and other students of British antiquities discovered the sweep and magnitude of their country's megalithic ruins, their attention turned to what little was known of ancient Britain. What they found, all but forgotten in a handful of Greek and Roman writings, were fragmentary references to a mysterious group of people called Druids.

Diviners, Poets, and Teachers

Who were the Druids? The honest answer is that we really don't know. Most of what was written about them in ancient times vanished forever when the Roman Empire collapsed in the fifth century of the

Common Era. All the surviving texts written about the Druids when they still existed add up to a total of ten pages or so in English translation.

This meager harvest offers little solid knowledge. Druids lived among Celtic tribes in Gaul (modern France), Britain, Ireland, and apparently nowhere else. Their name may have meant "wise ones of the oak," although scholars have suggested many other interpretations. They taught a secret wisdom that probably had something to do with trees, and their sacred places were groves in the forest. Some classical writers call Druids philosophers; others call them wizards. Not once do the sources call them priests, although this is how most archeologists interpret them today. Several sources divide them into the three categories of Bards, Ovates, and Druids, the Bards being poets, the Ovates diviners, and the Druids teachers.

Custom forbade writing down Druid teachings, and Druids-in-training had to memorize prodigious amounts of lore in verse form. Some students took twenty years to complete the course of study, which included theology, astronomy, and divination. After finishing their studies, Druids formed a privileged class, exempt from taxes and military service. They settled disputes and could part warring armies on the brink of battle.

Crucial questions about ancient Druids remain unanswered. Their origins? Julius Caesar, whose book on the Roman conquest of Gaul is the most detailed source we have, noted that Druid teachings were thought to come from Britain originally, while a Greek scholar claimed that the Druids got their lore from the Greek philosopher Pythagoras; no other writer refers to the subject. Their ceremonies? The unreliable Roman author Pliny describes Druids gathering mistletoe from an oak with a sickle of gold. Several others refer offhand to Druids playing some role when animals or humans were sacrificed to the gods. Most authors say nothing. Their organization? Caesar claimed that an Archdruid of Gaul was elected by other Druids, but no one else refers to anything of the kind. Their daily life? No ancient writer mentions it at all.

After the Roman conquest of Gaul and Britain, Druids fade from the classical sources. A few references from the third and fourth centuries suggest that they could still be found, but had come down in the world; one third-century Gaulish Druidess worked as an innkeeper. A few Latin texts from the early Middle Ages, mostly biographies of saints from the age of Christian conversions, refer to them but provide nothing new.

These scraps were all the scholars of Aubrey's time had on which to base their studies. Old Irish literature had more to say, but two centuries passed before anyone outside of Ireland realized that. Until then, ten pages of Greek and Roman references comprised nearly everything known about Druids. Folklore, tradition, and the mute evidence offered by the old stones and earthworks of a forgotten age provided other sources of uncertain value.

Yet it was enough. People all over Britain started studying Druids, writing about Druids, trying to tease out Druid secrets from any available source. A surprising number of them went on to become Druids, to embrace Druidry as a spiritual path in their own lives. The idea of a green way of wisdom, a spirituality rooted in nature and the living Earth, had a potent attraction for people who couldn't stomach either the rigid dogmatism of organized Christianity or the equally rigid nihilism of emerging modern science. Thus research into a long-forgotten tradition gave birth to a serious attempt to revive it.

The Rebirth of Druidry

According to traditions current in English Druid orders, a crucial step in this process took place on 28 November 1717, when a meeting of Druid enthusiasts at the Apple Tree Tavern in Covent Garden, London, founded the Ancient Druid Order, the first Druid organization of modern times. Some overly enthusiastic modern Druids portray this meeting as a grand event drawing white-robed Druids from far and wide. The reality must have been more prosaic: a dozen men in ordinary dress in a private room of a middle-class pub, taking the first uncertain steps toward a Druid revival between tankards of ale and cider.

To date, nobody has found solid documentary evidence of the 1717 meeting, but the tradition is plausible enough. Eighteenth-century England was bursting at the seams with societies, clubs, and secret or semisecret orders, most of them poorly documented. Even such famous organizations as the Freemasons, who founded their first Grand Lodge earlier that same year, preserved few records from this time. A Druid society could well have flourished in London at that time and left only equivocal traces of its existence. Still, whether or not the 1717 meeting happened, it's clear that, in these same years, the broader Druid Revival was under way.

The earliest modern Druids came to the Revival with diverse agendas. Some sought to rework Christianity into a greener form, while others rejected Christianity altogether and sought an alternative faith. Movements in this latter direction faced serious legal problems, since believing in more than one god and questioning the Christian doctrine of the Trinity were crimes under British law until the early nineteenth century. Druids with alternative views had to step warily. Nonetheless, some amazingly innovative religious thinking occurred.

The central theme of this new spirituality was pantheism—a belief that the universe itself is a living, divine being. The word *pantheism* was coined by the Irish writer John Toland (1670–1722), whose many interests included Druid lore. Toland is among those said to have attended the 1717 London meeting, and certainly his views on religion had a major impact on the Druid Revival. To pantheist Druids, nature was literally divine, the material body of God, and nothing natural was to be despised or rejected. In the society of the time, with its rigid formalities and moral codes, this was explosive stuff.

Yet the first century and a half of the Druid Revival also saw important links between Druidry and the Latitudinarian movement in the Church of England. Latitudinarians hoped to move the Church away from doctrinal squabbles, urged tolerance for dissent, and proposed a personal spirituality based on meditation and individual study. Latitudinarian Druids such as William Stukeley (1687–1765), whose Druid path led him to the Anglican ministry, held that Druid nature mysticism was entirely compatible with such a personal, non-dogmatic Christianity. Meanwhile, Druids of all kinds borrowed freely from the extensive Anglican literature on meditation and spiritual exercises.

These two currents—pantheism and Latitudinarian Christianity—pulled the early Druid Revival in different directions, but never quite pulled it apart. The two wings of the Revival sniped at each other, to be sure; Toland wrote a *History of the Druids* that was an extended satire on the Anglican Church, while Stukeley promoted Druid teachings among his fellow Anglicans as a way to counter Toland's pantheism. Both currents spoke to the needs of their time, however, and helped people see Druidry as a path for the present, not simply a curiosity from the past.

2

THE REVIVAL
TRADITIONS

THE SHEER LIMITATIONS OF the sources at hand forced Druidry to borrow and innovate from the very beginning. With so little to go on, the Revival squeezed the available evidence for whatever it would yield. Anything even slightly connected to the ancient Druids was fair game. An offhand comment by Caesar that Druids taught lore about the planets sent half a dozen Druid writers roaming Salisbury Plain, hoping to tease out secrets of Druid astronomy and cosmology from the megalithic sites there. Two ancient Greek authors compared Druids to Pythagoreans, a Greek sect devoted to sacred geometry and number mysticism; that was enough to put eager Druids in hot pursuit of Pythagorean lore, much of which ended up incorporated in Druid Revival teachings.

Even more important was the simple fact that the ancient Druids worshipped in groves and forest glades. This glimpse of woodland spirituality evolved into a potent theme of the Revival. Druids and forests fused so totally in the British imagination that a 1743 book on growing oak trees, among the earliest books of silviculture in English, was titled *The Modern Druid*.

All these patterns, the borrowing and innovation as well as the Arcadian vision, need to be understood in the context of the age. By the middle of the eighteenth century, the Industrial Revolution was under way, and its human and ecological costs were soaring out of sight. Mines, factories, and vast fetid slums sprawled across England. The urban poor, most of them country folk driven off land their ancestors had farmed for generations, faced a world in which children as young as eight had to work sixteen-hour days to keep their families from starving. In urban slums and factory districts, a new industrial landscape took shape, defined by red brick, black iron, and choking clouds of coal smoke.

This desecrated landscape gave new meaning to Druidry. In an age drunk with the power of an Earth-damaging technology, the vision of an older wisdom learned among trees forcefully challenged the legitimacy of the new industrial order. The second half of the eighteenth century, the first great flowering of Revival Druidry, was also the seedtime of modern ecological thought, the years in which British naturalist Rev. Gilbert White (1720–1793) wrote his pioneering *Natural History of Selbourne* and German polymath Johann Wolfgang Goethe (1749–1832) laid the foundations of a holistic science of nature. Insights such as these, and the broader movement of principled opposition to industrialism, played a crucial role in the growth of Druidry and helped shape its vision of a spirituality of green nature.

Historical scholarship also fed into the developing Druid tradition. Revival Druids made an eager audience for books about the Pagan past, comparative religion, myth, and folklore. From the beginning of the Revival, such works served as raw material for reconstructions of ancient Druid teachings. Most were far off the mark by today's standards, but were on the intellectual cutting edge of their own time. They also gave Druids tools that could be used to challenge the religious and social orthodoxies of the day.

Thus early nineteenth-century theories traced Paganism back to Noah via a hypothetical Helio-Arkite cult worshipping the Sun and Noah's ark. Druids borrowed these ideas, linked them to Welsh myth, wove them into their rituals and teachings, and proceeded to claim that, as good Helio-Arkites, their religion was older and purer than the Church of England's. By the last decades of the same century, scholars were arguing that all Pagan religions evolved from the worship of penis and vagina as symbols of life and fertility. Druids borrowed these theories and applied them to Christianity, proposing calmly that Jesus was a phallic symbol and churches were immense stone vaginas—all this in the middle of the Victorian era!

The deadpan humor of both of these examples was another trait that showed up early and often in the Revival. Thus William Stukeley poked fun at the wild theories circulating around Stonehenge in the early eighteenth century by proposing, with a straight face, that a tribe of intelligent elephants from Africa could have built the great stone circle. A century and a half later, Owen Morgan, Archdruid of the Druids of Pontypridd in Wales and a supporter of the sexual theory of religion, prefaced a book full of vivid erotic symbolism and blatantly sexual interpretations of Christian myth with a prim little note suggesting that only those with depraved imaginations would find anything offensive in its pages. Puckish comments of this sort kept Druidry's critics off balance, and kept Druidry itself from becoming too pompous.

Druids and Masons

Another current that flowed into the Revival came from Freemasonry. The Order of Free and Accepted Masons started out as a medieval guild of stonemasons in England and Scotland. In the seventeenth century, Masons' lodges began admitting "accepted Masons"—members from outside the building trades who were attracted by Masonic rituals and symbolism as well as the social side of membership. By the first years of the eighteenth century, accepted Masons formed a majority in most lodges, and Freemasonry transformed itself into a men's social club dedicated to self-improvement and charitable assistance, as it remains today.

Freemasons evolved an effective system of organization centered on local lodges and powerful ritual and symbolic tools for initiation ceremonies. During the eighteenth century, as Freemasonry became fashionable throughout Europe and the American colonies, their methods were borrowed by many other groups: the Odd Fellows, the Bucks, the Gormogons, the Mopses—and the Druids. It's probably not an accident that the traditional date of the Ancient Druid Order's origin in 1717 was a few months after the founding of the first Masonic Grand Lodge of England. Links between Druidry and Freemasonry branched out in many directions, and at least one significant Druid order, the Ancient and Archaeological Order of Druids (AAOD), founded in 1874 by the Mason and Rosicrucian Robert Wentworth Little, had close connections to esoteric circles within British Masonry.

Another aspect of Freemasonry helped drive these links with the Druid tradition. In the intensely class-conscious society of eighteenth-century Europe, some gentlemen Freemasons fretted about the awkward fact that their beloved order had started out as a working class trade union. Some Masons of the time chose to manufacture more romantic origins for their order. Connections between the Freemasons and crusading orders of knighthood such as the Knights Templar were widely alleged, despite a distinct lack of evidence, and of course the same allegations are repeated today.

Another attempt to find a pedigree for Freemasonry, however, traced it back to the ancient Druids. As with the Templar connection, lack of evidence was no barrier to eager Masonic imaginations. During the eighteenth and nineteenth centuries, many male Druids were also Freemasons. The two groups shared an interest in ancient mysteries and a dislike for dogmatic religion, and Druid orders with Masonic links, such as the AAOD, made contact easy. Thus Masonic traditions were caught up in the quest for surviving fragments of Druid lore. However inaccurate the results may have been historically, some Masonic material meshed well with Druid Revival teachings and gained a lasting place in the tradition.

Celtic sources, finally, played an important but complex role in the Revival. The ancient Druids survived in Ireland and Scotland long after they vanished elsewhere, and scraps of Druid lore can be traced to Irish and Scots bardic schools that still existed early in the Revival. Politics, however, made these sources too hot to handle for many years. The heirs of the exiled House of Stuart, driven from the British throne and widely hated in England for their attempts to abolish civil and religious liberties, drew on support from Ireland and the Highlands through most of the eighteenth century. This poisoned relations between English and Gaelic communities and made it impossible for English Druids to make use of Irish or Scots material until the threat of a Stuart reconquest was long past.

Welsh Druidry

Wales was another matter, however. While Wales became Christian as early as any country in Western Europe, remarkable survivals existed. Oxen were being sacrificed in half-Pagan rites at Clynnog Fawr as late as 1589. A few centuries earlier, in the last years of the Welsh struggle for independence, something very like a deliberate revival of Celtic Paganism seems to have taken place, setting mythic figures out of Welsh legend—Hu Gadarn, Ceridwen, Gwyn ap Nudd, and others— in place of the Trinity and saints of conventional Christian devotion. Thus the fifteenth-century bard Rhys Brydydd wrote:

> The smallest, if compared with small,
> Is the Mighty Hu, in the world's judgment,
> And he is the greatest, and Lord over us,
> And our God of mystery.

Such statements raised the hackles of the more orthodox clergyman-poet Sion Kent (d. 1420), who wrote:

> Two kinds of inspiration in good truth
> Exist and manifest their course on earth:
> Inspiration from sweet-spoken Christ,
> Orthodox and gladdening the soul,
> And that most unwise other inspiration,
> Concerned with false and filthy prophecy
> Received by the devotees of Hu,
> The unjustly usurping bards of Wales.

From this historical distance, it's hard to be sure just what was going on here. Some historians have suggested it may have been pure literary affectation. Still, the spluttering wrath of Sion Kent's poem suggests that something more serious was afoot. Kent's stark opposition between Christ and Hu, Rhys Brydydd's "god of mystery," is hard to explain in anything but a theological sense.

This shadowy medieval revival set the stage for the rise of Welsh Druidry in the nineteenth century. The Welsh poet and scholar Edward Williams (1747–1826), usually known by his bardic name Iolo Morganwg (Iolo of Glamorgan), launched this branch of the Revival with writings and public rituals based on lore allegedly passed down through Welsh bardic circles. The trickster of the Druid Revival, Iolo was not above giving out his own exquisite poetry as the work of medieval Welsh bards or slipping his own work into otherwise authentic texts. His writings are a mix of traditional Welsh lore, ideas borrowed from earlier Druid Revival authors, and his own brilliant inventions. For good or ill, no other individual had as great an impact on the nineteenth-century Revival.

Welsh *eisteddfodau* (bardic assemblies) adopted Iolo's rituals and symbolism in the early nineteenth century and launched an utterly respectable Christian branch of Druidry that remains active to this day. Yet Iolo's legacy also led in wilder directions. Following his lead, Myfyr Morganwg (Edward Davies) and his successor Morien (Owen Morgan) headed a Druid order at Pontypridd in Wales all through the late nineteenth and early twentieth centuries. The astonishing Dr. William Price—Druid, labor organizer, socialist, naturopathic physician, nudist, and proponent of cremation, among much else— had connections to Pontypridd's Druids as well. Price, the most colorful of the Welsh Druids of his day, used to chant "a song of the primitive bard to the Moon" at the rocking stone at Pontypridd, clad in a white tunic, green trousers, and red waistcoat, with the skin of a fox over his head and shoulders.

The nineteenth century was a difficult time, however, for connections between Druid Revival groups in England and France and their opposite numbers in Wales, Brittany, and other Celtic countries. The English government's oppressive policies in Ireland sparked a Celtic nationalist movement with little tolerance for people outside surviving Celtic cultures. Many Celtic nationalists turned to conservative

Christianity to counter an increasingly secular English culture, and even those who stayed interested in Druid lore rejected Druid groups outside the Celtic nations with considerable heat. Druid groups in England and elsewhere responded in various ways. Many continued to support causes such as Irish home rule, but others withdrew from the political sphere. The divisions in the Druid movement opened by the bitter politics of the nineteenth century remained unhealed throughout the century that followed, and are only just beginning to be resolved today.

The Druidry of the Revival

By the late nineteenth century, despite these troubles, Druidry evolved into a distinctive spiritual and cultural tradition, and became something of an institution in the British culture of the time. Druid groves (local groups), for the most part small and fiercely independent, pursued many variations on a common theme. Druid teachings varied from order to order and from Druid to Druid, but shared basic principles and traditions. Books, rituals, and symbols provided a language of ideas common to most Druids, but nothing like a Druid orthodoxy ever took shape—and few Druids would have tolerated one if it had.

Three main themes formed a broad consensus within which most Druids worked out their personal paths. First was a deep sense of reverence for the land. The image of the Druid sitting beneath a tree or a standing stone contemplating hidden mysteries, a commonplace of poets throughout the eighteenth century, became common reality in the nineteenth, although the white robes and pastoral staff of the poetic image generally gave way to the wool jacket and walking stick of the middle-class Victorian out for a day in the country. Druids were also at the cutting edge of movements in natural medicine and a healthier diet. A number of Druids from the late nineteenth century on were nudists as well, taking closeness to nature to its logical conclusion.

A second shared theme was personal spiritual development. Druid spirituality rejected orthodox notions of sin, judgment, and damnation and embraced instead a vision of human souls working out their personal salvation from life to life. Practices to foster inner development found a natural place in this vision. Methods of meditation borrowed mostly from Anglican sources formed the most widely used approach to inner development, though Druids borrowed freely from other traditions as well.

Public ritual formed a third theme that set Druidry apart from most other alternative traditions in Britain. From 1792, when Iolo Morganwg and a group of Druids including the poet William Blake celebrated the autumn equinox on Primrose Hill before a crowd of baffled and fascinated Londoners, Druid tradition affirmed that certain ceremonies were best done "in the face of the sun, the eye of light." Though it had its problems, public ritual had a spectacular advantage: it made the Druids familiar. Few people harbored dark suspicions about Druids, since anyone curious about them could walk up to the ritual circle, look at the billowing robes and ceremonial banners, watch the sword thrust into its sheath as peace was proclaimed to the quarters, and hear the words of the ritual. This gave Druids a respectability that helped keep them out of the scrapes suffered by more secretive traditions.

It's interesting, moreover, to compare these central themes with the core elements of other spiritual traditions then and now. Notably absent was any sense that Druids ought to hold any particular set of beliefs or restrict themselves to any particular set of practices. This freedom of thought and action ran straight through barriers few other people in the Western world could think of breaching. To most Victorians, for example, differences in religious beliefs opened up into vast social and cultural chasms. Yet as Druidry developed, it made space for an extraordinary diversity of religious faith. The pantheist and Latitudinarian views of the early Revival found common ground with a revived Pagan polytheism inspired by early British Pagans such as Thomas Taylor (1758–1835), and with Asian religions such as Buddhism and Hinduism.

Thus no one in the Druid community turned a hair when the Archdruid Myfyr Morganwg enlivened proceedings at an 1878 bardic gathering by invoking the Hindu goddess Kali. Most Druids, by this time, had come to believe that Druidry transcended limits of place, culture, or ethnic background. To them, the path they followed was a timeless spiritual possibility, open to all who shared its principles and were willing to learn its ways. In a world where strife between follow-ers of dogmatic religions was a daily event, many Druids placed high hopes in a spiritual path that built bridges instead of raising walls.

3

DRUIDRY IN THE WIDER WORLD

THE SPREAD OF DRUIDRY overseas helped foster this sense of its wider potential. Druid societies imported from Britain first appeared in America in the closing years of the Revolution, and by the First World War, most Druid orders had sizeable American and Canadian branches and presences Australia, New Zealand, and much of Europe.

French Druidry has a more complex history. Interest in the ancient Druids went back even further in France than in Britain, but the political power of the Catholic Church made it difficult to follow a Druid path openly until well into the nineteenth century. Once the barriers were down, however, French Druidry made up for lost time, rapidly creating an intensely creative Druid scene with its own distinct flavor and transmitting it to French-speaking regions in Canada. Brittany, the once independent Celtic region of France, contributed much to this development, although the pressures of competing nationalisms meant that Breton and French Druids had a complex and difficult relationship similar to that of Welsh and English Druids.

All these patterns remained in place through the twentieth century, although the Druid tradition continued to develop in new ways and drew on an ever-wider range of influences: Edwardian magical orders in the early years of the century, Asian spiritual traditions, the nature-oriented Woodcraft movement, the cultural revolution of the Sixties, and the Pagan renaissance in the century's last decades. Yet powerful factors forced a reassessment of 300 years of Druid tradition during these same years.

The rise of scientific archeology was the most important of these. As scholars made the past yield up its secrets, core assumptions of many Druid orders were overturned. The traditional image of the Druid, wise in the ways of green nature, gave way in scholarly literature to an "obscure barbarian priesthood" of no interest to anyone but specialists.[2]

This dismissal of Druidry, however, was neither as objective nor as disinterested as its promoters claimed. Archeology in the early twentieth century was a new science, eager to secure prestige and funding by staking out a territory and defending it against all comers. The Druid Revival, with its very different portrayal of the British past, could not be tolerated. Thus the archeologist Glyn Daniels wrote letters demanding that "those horrid bogus Druids" be barred from Stonehenge, while his colleague Stuart Piggott condemned "self-styled Druids which today represent the fag-end of the myth" as "pathetic."[3]

Despite mixed motives, however, archeologists and historians were able to show that the Druidry of the Revival was a modern spiritual movement, not an ancient one. Many Druids refused to accept this. By the early twentieth century, some Druid orders claimed unbroken lineages back to the ancient Druids. Such claims became hard to defend. Once made, however, they were even harder to give up. Thus even Druids as serious and intellectually capable as Ross Nichols (1902–1975), founder of the Order of Bards Ovates and Druids (OBOD) and a dominant figure in Druid circles in the decade or so before his death, clung to historically unlikely claims of a continuous tradition as if the spiritual validity of Druidry depended on them.

It took many years for Druids to realize that the historical roots of a tradition don't affect its spiritual validity—that Druidry's relevance depends on the gifts it has to offer here and now. As a living spiritual tradition with three centuries of achievements to its credit, Druidry can stand on its own and has no need of ancient lore to prop it up. By the dawn of the twenty-first century, the most influential Druid Revival orders had come to terms with their own history and abandoned inaccurate claims about the past.

Druidry Today

But what of the name "Druid" and the imagery that surrounds it? A case can probably be made for finding some other label for the tradition, but there are solid reasons against this. First is the sheer historical fact that the Druids of the Revival took that name, and not another; they were inspired by the ancient Druids and not some other ancient priesthood; and they have been known by that name ever since. For three centuries, the word "Druid" has meant, among other things, a participant in the Revival. Relabeling that movement "British Universalist Post-Anglican Latitudinarian Pantheist Neo-Pythagorean Nature Spirituality" or some other long-winded term is hardly an improvement.

Yet the Revival has another claim to the name Druid. Near the beginning of his book on Druids, Stuart Piggott asked:

> [W]hy has a priesthood within the barbarian pre-Roman Celtic religion, attested by a handful of some thirty scrappy references in Greek and Roman authors, many little-known and some downright obscure, come even to be remembered at all except by scholars nearly two thousand years after its official suppression by Roman authority?[4]

It's a fair question, and the answer is the Druid Revival. Ancient Druids are more than a footnote in history books today because a few visionaries in the early eighteenth century discovered them, were inspired by what they found, and set out on a quest to reconstruct a Druid spirituality in their own time. The innovative path they created put down deep roots in the Western world's imagination, and effectively defined the meaning of the word "Druid" for the modern world. Even those who most roundly reject the Druid Revival and all its works still commonly think of Druids in the terms the Revival created.

Thus when undergraduates at Carleton College, an Episcopalian college in Minnesota, decided in 1963 to protest a college requirement for Sunday chapel attendance, the "Druid church" they concocted had an astonishing resemblance to Druid Revival traditions, even though none of the culprits had any direct contact with traditional Druid organizations. The Reformed Druids of North America (RDNA), as they named their prank, had a thoroughly Druid sense of humor and a pantheist theology that John Toland would have heartily approved. The RDNA's statement of belief, in the mock-formal language favored by its founders, ran as follows:

1. The object of the search for religious truth, which is a universal and never-ending search, may be found through the Earth-Mother; which is Nature; but this is one way among many.

2. And great is the importance, which is a spiritual importance, of Nature, which is the Earth-Mother; for it is one of the objects of Creation, and with it do people live, yea, even as they do struggle through life are they come face to face with it.

For working purposes, these two tenets took on a simpler form:

1. Nature is good!
2. Likewise, Nature is good!

It would be hard to sum up the central theme of the Druid Revival more succinctly. Like the Revival itself, the RDNA went in unexpected directions. Although Carleton College's chapel attendance requirement was abolished in 1964, many RDNA members found that the liturgy and teachings invented for the "Druid church" was a meaningful spiritual path. The RDNA remains active to this day, and easily half the Druid organizations in North America today trace their origins to it.

Druidry and Modern Paganism

The extraordinary revival of Paganism throughout the Western world in the last decades of the twentieth century brought Druidry publicity and new members, but it also embroiled traditional Druid orders in occasional quarrels. Some newly minted Pagan groups tried to lay exclusive claim to Celtic tradition; others denounced the Druid Revival for its tolerance of monotheism, its willingness to borrow useful material from sources rejected by mainstream scholarship, and the inability of eighteenth- and nineteenth-century Druids to respond to late twentieth-century political causes in advance. Meanwhile, other Pagans borrowed freely from Druid Revival sources, and a number of traditional Druid orders returned the favor by borrowing from the new Paganism.

The unfolding ecological crisis of the late twentieth century brought even more attention to Druidry, and continues to do so today. After 300 years, the Industrial Revolution launched in seventeenth-century Britain now dominates the globe, with consequences that become more problematic with each passing year. Druidry, although born alongside the Industrial Revolution, traced out a radically different way of relating to the Earth. Its principled refusal to join industrial society's war against nature gave it a strong appeal to people seeking a saner way of life. In turn, many Druid organizations have taken up environmental causes with gusto, with tree-planting programs a particular favorite.

The results have gone mostly unnoticed by the media and mainstream culture, but they mark a sea change in Western religious history. Michel Raoult, one of the few scholars to study modern Druidry, estimated, in his 1980 dissertation *Les Druides: Les Sociétés Initiatiques Celtiques Contemporaines*, that Druid organizations worldwide boast 1 to 2 million members. That's an impressive figure for an alternative spiritual movement, and the community has added members steadily since then. If its growth continues, the modern Druid movement may yet become a significant factor in today's religious and cultural scene.

What will happen then? It's anyone's guess, since Druidry is an ongoing quest, shaped by the challenges and needs of each age. Still, the ecological crises and inner discords that threaten an increasingly fragile industrial society make Druidry at least as relevant today as it was when John Aubrey rode into the village of Avebury and discovered a riddle written in ancient stone.

Further Reading

A good history of the Druid Revival has yet to be written, and Druids interested in the origins and growth of their traditions will have plenty of digging to do. These are some useful sources:

Margot Adler, *Drawing Down the Moon* (New York: Penguin, 1997). This seminal study of American Paganism includes material on the Reformed Druids of North America and other Druid groups.

Peter Beresford-Ellis, *The Druids* (London: Eerdmans, 1995). The most ambitious recent attempt to pull the surviving evidence into a coherent picture of the ancient Druids—speculative in places but well researched.

Philip Carr-Gomm, *The Druid Way* (Shaftesbury, Dorset: Element, 1993); *The Elements of the Druid Tradition* (Shaftesbury, Dorset: Element, 1991); *The Druid Renaissance* (London: Thorsons, 1996). Three capable works on contemporary Druidry, written or edited by the current Chosen Chief of the Order of Bards Ovates and Druids— essential background for students of Druidry.

Edward Davies, *The Mythology and Rites of the British Druids* (London: J. Booth, 1809). An influential and important work from the nineteenth-century Druid movement—heavy going for the modern reader, but worth the slog.

Miranda Green, *The World of the Druids* (London: Thames and Hudson, 1997). The best general introduction to Druids in print, focusing on ancient Druids, but with an unusually fair discussion of the Druid Revival and modern Druid groups.

Leslie V. Grinsell, *Folklore of Prehistoric Sites in Britain* (London: David & Charles, 1976). A county-by-county survey of English, Scottish, Welsh, and Cornish folklore associated with standing stones, barrows, and other prehistoric ruins—full of remarkable lore.

Daniel Hansen, *American Druidism: A Guide to American Druid Groups* (Seattle, WA: Peanut Butter Publishing, 1995). A useful overview of the modern American Druid scene, with summaries of most of the currently active Druid groups.

Richard Hayman, *Riddles in Stone: Myth, Archaeology and the Ancient Britons* (London: Hambledon, 1997). A history of ideas and attitudes toward British antiquity from medieval times to the present, including most of the recent controversies. Some parts of the Druid Revival are covered.

S. K. Heninger, Jr., *Touches of Sweet Harmony: Pythagorean Cosmology and Renaissance Poetics* (San Marino, CA: Huntington Library, 1974). The best single overview of the Renaissance Pythagorean tradition, from which the Druid Revival borrowed sacred geometry and much more.

T. D. Kendrick, *The Druids* (New York: Barnes and Noble, 1966). Originally published in 1927, this remains a good introduction to the evidence concerning the ancient Druids.

Euan W. MacKie, *Science and Society in Prehistoric Britain* (London: Paul Elek, 1977). A valuable work on the social dimensions of megalithic culture, with a good deal of relevance to Druid origins.

John Matthews, ed., *The Bardic Source Book* (London: Blandford, 1998); *The Celtic Seers' Source Book* (London: Blandford, 1999); *The Druid Source Book* (London: Blandford, 1996). Three massive tomes full of source documents and studies covering bards, ovates, and Druids from ancient times to the present, with much Revival material.

Dillwyn Miles, *The Secret of the Bards of the Isle of Britain* (Llandebie, Wales: Gwasg Dinefwr, 1992). A history of the Welsh Gorsedd with useful information on Iolo Morganwg and his role in the Druid Revival.

Colin Murray, *Golden Section Order Broadsheets* (London: Cantata Organica Press, 1956–1977). Issued by the Golden Section Order, these broadsheets cover a wide range of Celtic and Druid symbolism and magic, and are well worth study if you can find copies.

T. Islwyn Nicholas, *A Welsh Heretic: Dr. William Price, Llantrisant* (London: Foyle's, 1941). Short biography of one of the most eccentric and entertaining nineteenth-century Druids.

Ross Nichols, *The Book of Druidry* (London: Thorsons, 1990). Written by the founder of the Order of Bards Ovates and Druids, a complex, difficult, and essential book that repays careful study.

A. L. Owen, *The Famous Druids* (Oxford: Clarendon Press, 1962). A good general study of Druids in English literature from the late 16th to the early nineteenth centuries, the seed bed of the Druid Revival.

Stuart Piggott, *The Druids* (London: Thames & Hudson, 1975). A good introduction to the archeological evidence about ancient Celtic religion, mixed with Piggott's often entertaining splutterings about the Druid Revival.

Also, *William Stukeley: An Eighteenth-Century Antiquary* (London: Thames & Hudson, 1985). A valuable study of one of the first major figures in the Druid tradition, much more balanced than the first edition.

Theodore Roszak, *Where the Wasteland Ends* (New York: Anchor, 1972). A widely acclaimed book on the inner dimensions of the ecological crisis, drawing extensively on the same English Romantic circles that fostered the Druid Revival. Essential.

John Williams ab Ithel, ed., *The Barddas of Iolo Morganwg* (York Beach, ME: Weiser, 2004). The core document of nineteenth-century Druidry, compiled from manuscripts left by Iolo Morganwg—essential reading though very much a mixed bag.

Dudley Wright, *Druidism, the Ancient Faith of Britain* (London: E.J. Burrow, 1924). A good general overview of ancient Druidry as the Revival interpreted it. Draws heavily on Davies' *Mythology and Rites*, but is much easier going.

PART II:

WISDOM OF THE STONE CIRCLE

Three Triads of Druid Philosophy

4

THE FIRST
TRIAD

THE LORE OF DRUIDRY DRAWS on many sources and takes many forms. Enigmatic fragments of ancient Druid lore, wisdom teachings from surviving Celtic cultures, borrowings from other spiritual traditions, insights from the world of living nature, and the mute testimony of green earth and standing stone shaped a foundation on which three centuries of Druids have built in many different ways. To present the results in any detail would fill volumes, and provide little help and much confusion to the novice Druid. For this reason I've focused the discussion of Druid teaching here on nine core concepts, suggesting some of their meanings and a few of their complexities.

Others could have been chosen, but these nine touch on most of the central themes of Druidry—sometimes in unepected ways. The following pages say more than a casual glance may notice. As William Blake said, "The wisest of the ancients considered what is not too explicit as the fittest for instruction."

These nine concepts are divided up, in keeping with Druid tradition, into three triads. The first triad of Druid teachings covered here focuses on three fundamental images, each of which is itself a triad: the Three Rays of Light, the Three Circles of Manifestation, and the three elements of Druid natural philosophy.

Three Rays of Light

Einigen the Giant, the first of all beings, beheld three rays of light descending from the heavens. Those three rays were also a word of three syllables, the true name of the god Celi, the hidden spirit of life that creates all things. In them was all the knowledge that ever was or is or will be. Beholding the rays, Einigen took three staves of rowan and carved all knowledge upon them, in letters of straight and slanted lines. But when others saw the staves, they misunderstood and worshipped the staves as gods, rather than learning the knowledge written upon them. So great was Einigen's grief and anger at this that he burst asunder and died.

When a year and a day had passed after Einigen's death, Menw son of Teirwaedd happened on the skull of Einigen, and saw that the three rowan staves had taken root inside it and were growing out of its mouth. Taking the staves, Menw learned to read the writing on them and became famous for his wisdom. From him, the lore of the rowan staves passed to the Gwyddoniaid—the ancient loremasters of the Celts—and ultimately from them to the Druids. Thus the knowledge that had once shone forth in three great rays of light, passed through many minds and hands, now forms the wisdom of the Druid tradition.

The tale of Einigen the Giant and the three rays of light is the origin myth of the Druid Revival. The rays themselves form the core symbol of Druidry, and appear constantly wherever Druid Revival teachings have left traces. They represent *Awen*, the heart of the Druid path.

What is Awen? In its most basic sense, it means spirit, inspiration, and illumination. Awen is the inner light that gives the mind the ability to reach beyond itself. It's Awen that turns writers of verse into poets and shows glimpses of the future to prophets and diviners.

Druids of the Revival studied the lore of Awen and related it to knowledge and experience from many sources. They knew that people of all religions experienced moments of illumination in which the world took on profound meaning. Some of these experiences, they knew, happened when people turned their attention outward to nature or the great spiritual powers of nature that human beings call gods. They knew that other experiences of the same kind took place when people turned their attention inward to the core of themselves.

These two modes of experience are symbolized by two ways of drawing the three rays of light. The first, the invoking form \ | /, represents the experience of Einigen—the descent of Awen from a divine source outside the self. The second, more commonly used evoking form / | \ represents the experience of Menw—the awakening of Awen within the individual soul through study and contemplation (see figure 1). The two forms together form an emblem called the *Tribann*.

FIGURE 1. THE THREE RAYS OF LIGHT.

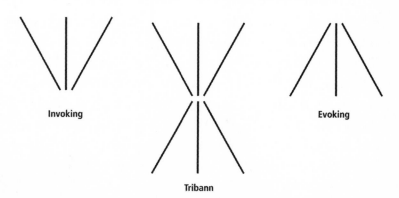

Invoking

Evoking

Tribann

The tale of the three rays of light is myth, not history. Yet the word "myth" needs to be understood with care. Too many people today use it to mean a story that isn't true. Real myths, however, aren't mere lies, even when they have nothing to do with historical fact. As the Greek philosopher Sallust put it: "Myths are things that never happened, but always are."[5]

Thus it matters not at all whether a person named Menw ap Teirwaedd existed, whether he came across the skull of a giant with three rowan shoots growing out of it, or whether this happened exactly 366 days after the giant's death. The daily-newspaper sense of literal fact that fundamentalists bring to sacred texts has no place in the Druid understanding of myth. The heart of any myth is not whether it happened—or, for that matter, who wrote it—but what it means and what it has to teach.

This insight itself is, in fact, among the things the tale of Einigen teaches. Einigen recognized the rays of light, and Menw recognized the staves, as bearers of knowledge. The others who saw the staves failed to grasp this, and worshipped the staves instead of learning from them. There's a potent lesson here on the difference between a religion of belief and a spirituality based on understanding. The stave worshippers treated the staves as objects of faith, and missed every teaching those objects were meant to convey. In the same way, someone can believe in the story of Einigen and Menw, and never grasp that their tale has something to teach.

This isn't the only lesson of the myth, however. The three rays are also three shouts and the three parts of the true name of the god Celi, the hidden source of all things. That name is the word Awen itself, divided into three syllables: Ah-Oh-En. Take a moment to think about what this means. Awen, the inner illumination that gives gifts of poetry and prophecy, is the true name of the source of all things. To touch Awen is to touch the divine energy that creates the universe. To write a poem that captures a flash of inspiration is to enact the deed of Einigen, whose rowan staves caught the wisdom shown forth in the three rays of Awen. To read a poem and catch a glimpse of the original inspiration is to enact the deed of Menw. Thus the tale of the three rays of light is told again in every work of creativity.

In Druid lore, the three rays are also the first rays of the rising Sun at the summer solstice, the winter solstice, and the equinoxes, shining on a standing stone and casting its shadow across the ground. The Sun's rays trace the invoking form, while the shadows trace the evoking form, as shown in figure 2.

FIGURE 2. THE THREE RAYS, THE SOLSTICES, AND THE EQUINOXES.

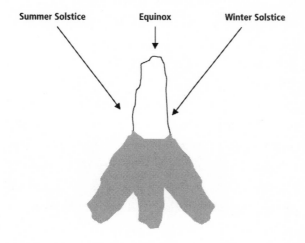

An ancient myth found in many lands describes the creation of the cosmos from the body of a primal giant. In these legends, the skull of the giant becomes the dome of the sky. The Welsh myth of Einigen, echoing these ancient tales, has rowan staves rising from the dead giant's skull in place of the dawns that define the year. Thus another meaning of the myth is enacted each year when the Sun rises at its three primary positions on the eastern horizon: southeast at midwinter, northeast at midsummer, due east at spring and fall.

Each ray has its own symbolism and meaning. The first, or midwinter, ray extends from the southeast \|/ or toward the northwest /|\. It is named Gwron ("goo-RON") and represents the Sun at its lowest point at the winter solstice, amid the darkness and cold of winter. Gwron is the knowledge of Awen.

The second, or midsummer, ray extends from the northeast \|/ or toward the southwest /|\. It is called Plennydd ("PLEN-nith") and represents the Sun at the midsummer height of its power. Plennydd is the power of Awen.

The third, or equinoctial, ray extends from the east \|/ or toward the west /|\. It is called Alawn ("AH-lon") and represents the Sun at the point of balance, midway along its rising and falling course through the cycles of the year. Alawn is the peace of Awen.

Gwron, Plennydd, and Alawn are also the names of three mysterious figures of legend, the three first bards of the island of Britain. An account from *Barddas* claims that they first set out the ranks of Ovate, Bard, and Druid among the Gwyddoniaid, the loremasters of the Celts, in the days before their wanderings brought them to Britain. Thus the three classes of Druids also correspond to the three rays. Bards, who preserve knowledge of the past, correspond to the midwinter ray. Ovates, who encounter the living power of the spiritual realms, correspond to the midsummer ray. Druids, who unite knowledge and power, and traditionally separated armies on the edge of battle to bring peace, correspond to the central ray of the equinoxes.

Three Circles of Manifestation

Not all Druid teachings take the form of mythic tales. Philosophy and formal instruction also have their place, as in another core element of Druid tradition—the three circles of manifestation. These three circles

—called Abred, Gwynfydd, and Ceugant—map out the Druid vision
of the origin and destiny of the human soul (see figure 3).

FIGURE 3. THE THREE CIRCLES OF MANIFESTATION.

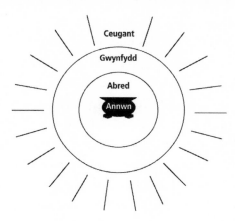

Abred (pronounced "AH-bred"), the lowest circle, is the beginning of
the soul's pilgrimage, the realm of plant and animal life in the world
of nature. Each soul comes from the Cauldron of Annwn (pronounced
"ah-NOON"), the mineral realm, which is also the reservoir of
unformed soul-stuff. Newly formed souls incarnate in the simplest
single-celled forms of life. As they master the lessons these have to
teach, they rise to more complex states. Life after life, form after
form, each soul gathers knowledge through countless reincarnations.

Druid teachings do not romanticize Abred. As Druids know, most
plants and animals lead short and difficult lives that end in painful
deaths. Plants struggle with one another for light and soil, and every
animal survives by killing other living things. None of this is accidental or unnecessary. A bardic triad teaches that each soul must see all
things, know all things, and suffer all things before passing beyond the
circle of Abred.

Human life marks the upper limit of Abred. Here, a new factor—
freedom—enters the picture. Animals and plants are living, conscious

beings, but instinct and biological drives bind them to the hard paths of Abred. Soul, at the human level, grows past these limits and can shape destiny for good or ill. Souls can pass beyond Abred to the circle of Gwynfydd, remain human, or fall back down the winding paths of Abred—possibly all the way to Annwn.

Tradition has it that each human failing copies the natural behavior of some animal or plant, and a human who can't outgrow a behavior is reincarnated as that animal or plant—not as a punishment, but simply as a means of finishing the learning process. Someone who displays the vanity of a cat or the empty-headedness of a sheep clearly didn't learn the lessons those forms teach, and must go back to relearn them, in the same way that someone who fails a math test may be assigned a remedial class. The creature that echoes your personal failings is called your *cydfil* (pronounced "KUD-vill"), and a skilled seer can glimpse a person's cydfil near them, as the shadow of a future incarnation. Each animal life is followed by a new ascent toward the human level, for Druid teachings insist that all beings will reach the second circle of Gwynfydd eventually.

Gwynfydd (pronounced "GWIN-vuth") is the fulfillment of the soul's pilgrimage. Gifts of knowledge, power, and peace won during the long journey through Abred unfold step by step in Gwynfydd as souls remember their journey up from Annwn and draw upon lessons learned along the way. Each soul learns to express its unique Awen, memories, and perception of the universe—the three things that set it apart from all other beings. In the process, each soul begins its journey through the circling paths of Gwynfydd.

This last concept sets Gwynfydd apart from more conventional ideas of the afterlife, for Gwynfydd is not a resting place, but the start of a new journey. A soul just arrived in Gwynfydd from the human level compares to the mighty ones of Gwynfydd as a single-celled organism at the lowest circle of Abred compares to a human being. Just as the soul of the single-celled organism will reach human form sooner or later, the soul new to Gwynfydd can expect to stand among the mighty someday. Yet Gwynfydd itself is not the highest circle.

Ceugant (pronounced "KYE-gant") is the third and final circle of manifestation, extending forever above Gwynfydd. You can think of Ceugant as infinity itself. Just as a traveler on an open plain can chase the horizon and never run the least risk of catching up to it, souls in Gwynfydd rise forever in knowledge, power, and peace and always find Ceugant beyond them. Ceugant represents the limitless potential that souls rising from Annwn have not yet realized. According to traditional Druid lore, the divine powers of the cosmos dwell in its eternal silences.

Like the three rays of light, the three circles of manifestation entered the mainstream of the Druid tradition by way of Iolo Morganwg's *Barddas*, although they weren't his inventions. They first appeared in the writings of Scots poet and Druid James Thomson (1700–1748) before Iolo was born. Did Iolo borrow the circles from Thomson, or did both take them from some other source? No one knows.

The three circles of manifestation can be taken quite literally, of course. Many Druids, in fact, do take them this way, as a useful way to understand the origins and destiny of the human soul. Seen in these terms, the concept has much to offer. The hope of life after death can call on plenty of evidence—near-death experiences, encounters with spirits, and children who remember details of previous lives—to back up the idea that human souls survive physical death and are reborn after an interval into new bodies. Mystics and philosophers have suggested for centuries that human beings have potentials that go far beyond anything we can imagine, and there's no good reason to think that humanity in its present form is as far as the process of evolution can go. The concept of the three circles makes sense of all these points in a straightforward and elegant way.

Yet the three circles also apply within the limits of a single life, or, for that matter, a single moment. We constantly face the choice between Abred and Gwynfydd, between the habits of the past and the potential for free choice that shapes the future. Whether or not our future lives depend on the thoughtful use of freedom, our present lives from day to day certainly do.

In fact, the three circles appear at every moment. Abred is the past, fixed in time and impossible to change. Gwynfydd is the present, opening onto freedom and infinite possibility. Ceugant is the future, the horizon that recedes before us as we advance. One lesson here is that it's unwise to get stuck in the past, and unproductive to strain forward to a future that's always out of reach. The realm of human freedom and potential is always the present.

These points can be traced out in more detail. Choices made in the past have power over the present, and this influence is traditionally called fate. Fate presses toward the repetition of past patterns. When you find yourself repeating the same experiences or can't seem to get past the consequences of some past event, fate is shaping your life. From a wider perspective, fate is the power of Abred, the presence in human life of the same rigid unthinking patterns that shape the lives of animals and plants.

Fate doesn't dominate all of life, however. The future also has power in the present, for each soul has unique gifts that seek expression. Their force is traditionally called destiny, the power that pulls the soul toward situations in which its potential can be fulfilled. When you find yourself thrust willy-nilly onto unfamiliar ground, faced with challenges you've never met before, and forced back on resources you didn't know you had, destiny is shaping your life. From a wider perspective, destiny is the power of Ceugant, the presence of the infinite and unknown.

Caught between fate and destiny, the human soul has a power of its own—the ability to choose. This ability is will, and it holds the balance between the pressure of the past and the pull of the future. When you act in the world instead of simply reacting to it, guiding your life by your own choices rather than being pushed and pulled by forces outside you, will is shaping your life. From a wider perspective, will is the power of Gwynfydd, the capacity for freedom and creativity in every moment.

Fate, will, and destiny weave the fabric of each person's life. Any two, however, can overpower the third. Thus, when will moves in harmony with destiny, the influence of the past has no power. When will follows the bidding of fate, the deeper potentials of the self remain asleep. When fate and destiny move together, will struggles against them in vain. When all three pull in different directions, the soul stumbles through life and achieves little. When all three work in the same direction, Finally, nothing in the universe can stand against them except the walls of Ceugant itself.

Families, cities, nations, and cultures have their own equivalents of fate, will, and destiny, and so does humanity as a whole. To read history is to watch these patterns unfolding on a grand scale, and the great crises of the present time offer extraordinary opportunities to watch the three factors at work.

Three Elements of Nature

Most people today, when they hear the word "elements," think of the ninety-two chemical elements known to modern science. The same textbooks that teach schoolchildren about these elements speak scornfully of an older belief in only four elements: Fire, Water, Air, and Earth. Yet that scorn is misplaced. Scientists in the eighteenth century took over the word "element," but changed its meaning. In fact, the four elements of the old system weren't elements in the same sense as hydrogen, boron, or manganese. In the older sense, elements are basic categories of human experience, not just material substances. Anger, for example, is Fire in the world of the emotions, and summer is the fiery season. To gauge the difference between ancient and modern elements, you might try finding an emotion or a season for helium!

One crucial difference hidden by this shift in meaning is that elemental systems in the old sense need not be mutually exclusive. There's apparently only one valid way to sort out atoms by their atomic numbers, but elemental systems of the older kind can be shaped as needed to respond to different situations. Thus many Western mystics and magicians expanded the four elements by adding a fifth called Ether or Spirit. Chinese philosophers used a different set of five elements: Wood, Metal, Fire, Water, and Earth. Each system has its own strengths and weaknesses, and many traditions use more than one.

Druid Revival lore contains an elemental system of its own: a set of three elements that first appears in Iolo Morganwg's early writings. Whether it's an invention of Iolo's or a scrap of some older teaching is anyone's guess, but the three elements have been part of Druid Revival lore ever since his time. These elements are *nwyfre*, *gwyar*, and *calas*.

Nwyfre (pronounced "NOOiv-ruh") is an old Welsh term meaning "sky" or "heaven." As an element, nwyfre is the source of life and consciousness, and modern Druids often refer to it simply as the life force. In some texts, *nef*, the modern Welsh word for "heaven," is used for nwyfre. Its image in nature is blue sky.

Gwyar (pronounced "GOO-yar") literally means "blood" in old Welsh, but its more general meaning is "flow" or "fluidity." As an element, gwyar is the source of change, motion, growth, and decay. In some texts, the word *dwr*, the modern Welsh word for "water," is used for gwyar. Its image in nature is running water.

Calas (pronounced "CAH-lass") comes from the same root as *caled*, Welsh for "hard." It means "solidity." As an element, calas is the source of form, differentiation, manifestation, and stability. In some texts, the word *daiar*, the modern Welsh word for "earth," is used for calas. Its image in nature is stone.

In one sense, these three elements are simply three parts of the more common four-element system, with nwyfre equal to Air, gwyar to Water, and calas to Earth. As a system in their own right, however, they offer a complete vision of the universe and have important lessons to teach. According to Druid philosophy, everything in the universe is made up of these three elements in some combination, with one being dominant. All are forms of a primal substance that Druids call *manred*.

If you haven't learned to think in elemental terms, it may take some work to get your mind around the three Druid elements. Take a moment to mull over some examples. The food on your dinner table is a good example. The calas element of your dinner is the raw material: meat, grains, vegetables, fruit. The gwyar element is the cooking process that turns the raw material into the meal on your plate. The nwyfre element is the mental dimension: the selection of ingredients, the choice of recipes, and the skill and personality of the cook.

It's not hard to find other examples, but you may wonder what value such exercises have. The system of chemical elements has obvious powers. Scientists use it to shape matter in a galaxy of ways— some helpful, others less so. Can anything as useful be done with the elements of Iolo's Druid philosophy, or with the four or five elements of the broader tradition?

Nwyfre, gwyar, and calas have little relevance to physics or chemistry, to be sure. Their value lies elsewhere. Like other traditional elemental systems, the three Druid elements make sense of patterns throughout the universe of our experience. As tools for thinking, their power lies in their ability to point the mind toward insights and sidestep common mistakes.

Take the common modern habit of thinking about the universe purely in terms of physical matter and energy. This works well when applied in certain ways, but it works very badly when applied to human beings and other living things. Time and again, well-intentioned experts using the best scientific tools have tried to tackle problems outside the laboratory and failed abjectly. Rational architecture and urban planning, scientific agriculture and forestry, and logical schemes for education and social reform often cause more problems than they solve and fail to yield the results predicted by theory. Why? The theoreticians thought only of gwyar and calas, change and stability, energy and matter. They left something out of the equation: nwyfre, the subtle element of life, feeling, and awareness. They forgot that any change they made would cause living things to respond creatively with changes of their own.

All three elements must be considered in every situation. What is the thing in question, what does it do, and what does it mean? What will stay the same, what will change, and what will respond to the change with changes of its own? This sort of thinking is one of the secrets of the Druid elements. Table 1 gives some of the primary correspondences.

TABLE 1. THE THREE DRUID ELEMENTS			
Element	Nwyfre	Gwyar	Calas
Symbol	Sky	Water	Stone
Color	White	Blue	Green
Kindred	Gods	Ancestors	Spirits
Hallow	Fire	Well	Tree
Circle	Ceugant	Gwynfydd	Abred
Level of Being	Spirit	Energy	Matter
Level of World	Overworld	Underworld	Midworld
Realm of Midworld	Sky	Sea	Land
Part of Self	Mind	Life	Body
Part of Time	Future	Past	Present
Station of Sun	Summer	Spring/Fall	Winter
Station of Moon	Full	Half	New
Druid Tool	Wand	Cauldron	Crane Bag
Branch of Druidry	Druid	Bard	Ovate

The three elements make a particularly useful tool for thinking nowadays because of a habit of mind deeply entrenched in modern culture. Many people today divide every situation into two, and only two, factors. This by itself isn't necessarily a problem, but often the two factors are portrayed as absolute opposites with no common ground uniting them, and this leads to trouble. Worse trouble occurs when the opposites are assigned moral labels—one seen as completely good and the other as absolutely evil. Think of political and religious squabbles in recent decades and you'll find more examples than you can count, each one full of this sort of dualistic thinking: those who aren't with us are against us; you're either part of the solution or part of the problem, and so on endlessly.

Back in the nineteenth century, schools of esoteric philosophy allied to French Druidry worked out ways to overcome this habit of mental dualism. In this teaching, any division into two is called a *binary*, or more fully, an unresolved binary. Binaries make useful tools for thinking when you need to focus on differences, but they produce a distorted picture unless they're balanced by something else. The opposite distortion comes from a *unary*, a view that considers only one factor and focuses attention exclusively on similarities.

The first number that guides thinking into balance is the number three. Divisions into three are called *ternaries*. Every ternary, according to this teaching, consists of two things opposed to each other, and a third that connects them. Thinking in ternaries considers both differences and similarities. While it's not foolproof, it thus sidesteps some common obstacles to understanding.

Problems that can't be solved in binary thinking often find ready solutions once a third factor comes into play. Finding this third factor was, therefore, a common training exercise in these traditions. Students were presented with opposing views from philosophy, politics, and daily life, and told to find the third point of view that resolved them into unity—or, in the language of the system, to neutralize the binary. In this way, they also learned to see how the two sides of a binary support, define, and require each other.

Ternary thinking has limitations of its own, however, that can be avoided by using other numbers to think with. Every number between one and thirteen, according to Druid teaching, has its own special logic and usefulness. Yet the logic of the ternary has a special meaning to Druids. It helps counter the pervasive binary thinking of modern culture, and also resonates powerfully with the teachings of the Druid Revival. There's a deeper connection still, for ternaries appear again and again in traditional Celtic myths, legends, and folklore. It's not an accident that Welsh and Irish bards assembled their lore in the form of triads, or that love triangles, triple quests, and threefold deaths provide the framework for so many Celtic tales.

Ternary thinking solves problems created by too much reliance on binaries. For this reason, cultures with threefold patterns offer glimpses of a more balanced way of life to those caught in a harshly dualistic society. This is one of the reasons why Celtic cultures have such a strong appeal for many modern people. Yet this can also become a trap if the relationship between binary and ternary is treated itself as a binary. Drawing a binary distinction between binaries and ternaries (or between modern society and another culture), picturing them as irreconcilable opposites and labeling them in terms of right and wrong, echoes the modern world's habitual thinking while pretending to escape from it. What third factor might neutralize this binary?

Thus, while ternary symbolism has a central role in many of today's Druid traditions, few rely on it exclusively. Instead, many different numbers weave in and out of modern Druidry. Each has lessons to teach, and different grades of initiation focus on numbers that express their specific way of working with nature spirit. Similarly, the three Druid elements are not the only elements Druids use. The four elements Earth, Water, Air, and Fire have an important part to play, and the fivefold system, with Spirit at the center, also has a place.

These systems of three, four, and five elements have subtle and important connections. The three Druid elements become four when fire, or *ufel* (pronounced "IV-el") in old Welsh, is added. (The word *tan*, the modern word for "fire" in Welsh, is also used in some sources.) The four elements correspond to the four directions, seasons, times of day, and much more, and they play a primary role in many traditions of Druid ritual (see Table 2). The four Alban Gates—the solstices and equinoxes, the traditional Druid holy days through the eighteenth and nineteenth centuries—correspond to the four elements, as do the symbols on the altar in a grove of the Ancient Order of Druids in America, and the four levels of AODA initiation.

TABLE 2. CORRESPONDENCES OF A FOUR-ELEMENT SYSTEM				
Name	Nwyfre	Ufel	Gwyar	Calas
Symbol	Air	Fire	Water	Earth
Color	Yellow	Red	Blue	Green
Direction	East	South	West	North
Season	Spring	Summer	Autumn	Winter
Alban Gate	Eiler	Heruin	Elued	Arthuan
Time	Dawn	Noon	Dusk	Midnight
State of Matter	Gas	Energy	Liquid	Solid
Altar Symbols	Incense	Lamp	Water	Salt
AODA initiations	Candidate	Apprentice	Companion	Adept

The four elements, in turn, become five when Air, or *awyr* (pronounced "OW-ir") in Welsh, is added. (The words *wybr*, "sky," and *ffun*, "breath," are sometimes used for air in the old sources.) This may seem like a paradox, since nwyfre is already air! Yet in most of the world's languages, words for "air" and "breath" and "spirit" have common roots: the English word "spirit" itself comes from a Latin word for "breath," which also appears in words such as "respiration." As with so much in Druidry, a rigid approach has little value here, and it's best simply to recognize that Air and nwyfre fuse together when you're working with three or four elements, but draw apart when you work with five.

The system of five elements is central to the modern Druid use of Ogham, an ancient Irish alphabet (see chapter 5). The Ogham alphabet consists of four groups, or *aicme*, of five letters each, along with five additional letters, or *forfedha*. Each of these sets of five corresponds to the five elements, among other things. From this correspondence, key elements of Druid practice unfold (see Table 3).

TABLE 3. CORRESPONDENCES OF A FIVE-ELEMENT SYSTEM					
Name	Nwyfre	Ufel	Awyr	Gwyar	Calas
Symbol	Spirit	Fire	Air	Water	Earth
Color	White	Red	Yellow	Blue	Green
Direction	Center	South	East	West	North
Ogham Aicme	Forfedha	Second	First	Third	Fourth
First Aicme	Beith	Luis	Nuin	Fearn	Saille
Second Aicme	Huath	Duir	Tinne	Coll	Quert
Third Aicme	Muin	Gort	Ngetal	Straif	RuisIoho
Fourth Aicme	Ailm	Onn	Ur	Eadha	Mor
Forfedha	Koad	Oir	Uilleand	Phagos	

5

THE SECOND
TRIAD

THE SECOND TRIAD OF DRUID teachings explored here moves into the realm of spiritual and magical symbolism. Here the triadic system spins apart into a dance of other numbers. The solar and telluric currents form a twofold pattern of primary energies; the eight stations of the Druid year provide a basic cosmological pattern; the twenty-five letters of the Ogham offer an alphabet of concepts and energies. From their interplay unfolds the organic pattern of the Druid cosmos.

Above and Below

While the modern obsession with twofold patterns may be a source of problems, twofold patterns themselves can be useful when approached in a balanced way, and they have a valid place in Druid teachings. One such pattern defines sources of nwyfre drawn on by Druids in ritual and meditation: above, linked to Sun and sky, the symbols of nwyfre itself; and below, linked to water and stone, the symbols of gwyar and calas.

To understand what this implies, you must know a little more about nwyfre. If you've watched the Star Wars™ movies, you know most of what you need to know already; just think of nwyfre as the Force. George Lucas, creator of the Star Wars™ saga, borrowed the concept from Japanese martial arts, in which nwyfre is called *ki*. He could have found the same idea almost anywhere, for the concept is all but universal. The same force is called *prana* in India, *qi* in China, *n:um* (the : is a click) among the Kalahari bushmen, *ruach* in Hebrew, and *pneuma* in ancient Greek and early Christian writings. Today's Western industrial societies seem to be the only cultures in human history that lack a common word for nwyfre.

Like the Force in the Star Wars™ movies, nwyfre is everywhere, but more concentrated in some things than in others. The two most intense concentrations of nwyfre within our reach are in the Sun on the one hand, and deep within the Earth on the other. The two aren't interchangeable, however, for, like water, nwyfre takes on the qualities of the things it passes through. Nwyfre from the Sun and nwyfre from the Earth thus have different properties and powers.

Nwyfre from the Sun fills Earth's upper atmosphere and moves downward. It's stronger when the Sun is above the horizon and strongest of all at dawn and noon, but present even at midnight. At higher elevations, it has more force, which gives mountain air its healing properties and leads monks and hermits to choose high mountain country for their retreats. This mode of nwyfre is called the solar current. Its traditional mythic symbols are birds such as the eagle, the hawk, and the heron; a technical term for it in occult writings is *aud* or *od*.

Nwyfre from the Earth moves upward through stone and soil. It's strong around green plants, especially trees, and wherever water flows from beneath the Earth; holy wells and springs blaze with it. It follows complex cycles in time and space, shaped by lunar phases, planetary and seasonal changes, and the subtle interplay of underground water with stone and soil. This mode of nwyfre is traditionally called the telluric current (from Tellus, an old name for the Earth). In the language of alchemy, it's named the mercurial current and the secret fire. Its mythic symbols are the serpent and the dragon; occult lore calls it *aub* or *ob*.

These two currents of energy provide the warp and weft from which Druids weave their magic. In more than the obvious sense, they move in opposite directions. The solar current stimulates energy centers in the chest, throat, and hands, awakening intellect, abstract emotion, and impersonal awareness. It is symbolically masculine and can be called the current of wisdom. The telluric current stimulates energy centers in the belly, pelvic basin, and feet, engendering life, love, passionate energy, and vital force. It is symbolically feminine and can be called the current of power. The two currents relate as Shiva and Shakti do in Hindu Tantric lore. Shiva is absolute awareness; Shakti is absolute energy. Without Shakti, Shiva is empty consciousness; without Shiva, Shakti is blind force. Together, they create the universe.

It's surprisingly easy to lose sight of this latter point. Too many forms of spirituality exalt the solar current and its effects on the human energy system, while condemning the telluric current and its effects. In extreme cases, the solar current is identified with the divine and all goodness, the telluric with the demonic and all evil. Inevitably, some people rebel against this by reversing the interpretations, as though it were possible to stop being dualistic by standing dualism on its head. What's missing is the realization that either current alone is barren. Only when the two flow together does creation follow.

Each human being thus finds a balance between solar and telluric energies. Despite the symbolism, physical gender has little to do with this. Every human being partipates in both energies. Men and women alike range from primarily solar to primarily telluric personalities, although there's a slight tendency for women to express the telluric current more, while men tend to be more centered in the solar. Popular culture in the modern Western world echoes this in an inverse and destructive way, with gender stereotypes that entice women to surrender their power and men to surrender their wisdom.

In this sense, then, individual human beings make the third factor that neutralizes the binary between solar and telluric currents. Indeed, any living thing can fill the same role. Yet the currents also combine to produce a third current, the central secret of Druid magic.

This third current is called *aur* and *or* in occult writings. Some writers call it the Green Ray, although this term has other meanings. Here, we will call it the lunar current. Its mythic symbols include the egg, the jewel, and the child. It mediates between solar and telluric currents in the same way that the Moon mediates between Sun and Earth. Generated in an individual, it awakens the highest energy centers in the midbrain and the crown of the head, bringing healing, creativity, expanded awareness, and illumination. When it's generated on a larger scale and pours out across the land, it brings fertility, abundance, and peace.

The solar and telluric currents are primary and always present. They existed before the first living things arose on Earth, and will exist as long as Sun and Earth. The lunar current, by contrast, depends on the two primary currents, just as the Moon depends on the Sun for its light and the Earth for the stability of its orbit. Furthermore, the lunar current must be created out of the fusion of solar and telluric currents. It doesn't happen by accident.

The methods by which the lunar current is created are among the innermost teachings of Druidry, and of many other traditions as well. Yet the whole art can be learned by watching a seed sprout and grow into a tree. The newly sprouted seed brings sunlight and soil together and transforms them into the substance of the growing plant. It also brings solar and telluric currents together to form a spark of the lunar current, or Green Ray, at its meristem, the point of new growth. In the fusion of above and below, two become three and a new force is brought into play. All this ties into a symbolism present in one of the world's most commonly read and rarely understood legends, the legend of the Grail.

What is the Grail? Traditionally this was one of three questions asked by questing knights at the Grail castle. The Grail romances don't record the answer, but one possibility answer can be guessed from what we've already discovered. Each current has a symbolic form: the telluric current, or secret fire, is represented by a fire-triangle with one point upward; the solar current appears as a disk, like the Sun; the lunar current is symbolized by a crescent or half-moon shape open toward the top. Combine the three shapes and a familiar image appears.

**FIGURE 4. THE SYMBOLS OF THE CURRENTS
AND THE GRAIL.**

The relevance of this bit of imagery reaches deep into the Grail legend itself, and into Druid magic as well. In the Grail romances, the wounded Fisher King and the withered land could not save themselves from the terrible curse that bound them. They had to wait for a knight to vanquish the enchantments surrounding the Grail castle and ask the unasked question.

The kingdom of the Grail is the individual human being; on a wider scale, it's also the living Earth. Indeed, human greed and folly in the modern world make the Earth more and more like the Grailless lands of legend with each passing year. As nature withers, our inner lives wither in response. Living in an environment of machines and barren concrete, it's no wonder so many people have become barren and mechanical within. The wonder is that some still keep a spark of the green flame alive—a spark that someday must kindle a mighty blaze, once we make our way through the enchantments of our own dysfunctional society and ask the unasked questions that surround us at every moment.

Visionary scholars have described an age long past when sages had mastered the secrets of the three currents, when the telluric current flowed through lines traced on the landscape to sacred hills, then leaped down prepared channels to stone circles nearby where it fused with the solar current and sent the secret fire of the lunar current surging outward to bless the land. Whether or not this is an accurate image of the past, the present and future call us to work green magic of a similar kind to heal our world. This call flashes through the heart of the Druid path like moonlight through the bare branches of winter.

Stations of the Year

The polarity between above and below—between solar and telluric currents—defines a vertical axis around which Druidry turns. The circle Druidry traces around this axis takes its definition from other patterns, however. Time rather than space, and processes of change rather than states of energy, provide the framework for the eightfold year.

The eightfold year took many centuries to develop, and didn't take its present shape until modern times. Early records in Britain and elsewhere show a complex, localized patchwork of calendar customs in which no one pattern stands out. From the dawn of the Druid Revival, however, the rising midsummer Sun at Stonehenge provided a core image of the tradition. For many years, Druids celebrated the solstices and equinoxes—the points of the year most clearly marked by the old stone circles—as the four Alban Gates. This cycle is still the basic minimum in many traditional Druid orders, but most Druids today use a more complex calendar instead.

Table 4 gives the eightfold year of the modern Pagan tradition. Many people believe that this calendar dates from ancient times, but it didn't actually take its current form until the early 1950s, when two English Druids, Ross Nichols and Gerald Gardner, combined the old fourfold Druid calendar with the "cross-quarter" days—Samhuinn, Imbolc, Belteinne, and Lughnasadh—from medieval Irish texts. The fusion worked magnificently, producing a sacred calendar that catches the rhythms of the natural world in a balanced cycle around the wheel of the seasons.

Gardner, who was the creator of modern Wicca as well as a Druid of the Ancient Druid Order, brought the new calendar into Wicca as well. Unfortunately, like Iolo Morganwg and many others before him, Gardner found it necessary or useful to make up stories about the origin of his teachings, inventing ancient sources for the extraordinary product of his own creative genius and his borrowings from many contemporary traditions. Thus, generations of Wiccans and other Pagans, faithfully handing on teachings as they received them, have unwittingly spread inaccurate ideas about the calendar's origin and history.

TABLE 4. THE EIGHTFOLD YEAR		
Name	**Pronounced**	**Approximate Date**
Samhuinn	SOW-wen	1 November
Alban Arthuan	AHL-ban ARTH-u-an	21 December
Imbolc	IMM-olk	2 February
Alban Eiler	AHL-ban EYE-ler	21 March
Belteinne	BELL-chin-nuh	1 May
Alban Heruin	AHL-ban HEH-ru-in	21 June
Lughnasadh	LOO-nah-suh	1 August
Alban Elued	AHL-ban ELL-u-ed	22 September

Yet it bears repeating that the age of a tradition has no bearing at all on its relevance or value. Half a century of experience has shown that the eightfold year makes a perfect seasonal cycle of festivals for a nature-centered spirituality. Each station of the cycle has a counterpart on the other side of the wheel of time: winter balances summer, spring balances fall, harvest balances planting, life balances death, all reflecting the balance of the world of nature. A new festival comes every six and a half weeks on average, far enough apart so that each has its own seasonal character, close enough so that every point in the cycle relates to the others. It's an elegant conception, and well deserves the importance it has gained in modern Druidry and the modern Pagan revival as a whole.

Despite the popularity of the eightfold year, many of its potentials have received little attention in the half century since it came into general use. The best way to approach those potentials is to move around to another side of the question and consider the role of gods and goddesses in the wheel of the year.

What are the gods? Again, ask any three Druids and you'll get at least six answers. Theories about the nature of the gods are much discussed in some Druid circles, but don't have the importance they have in some other spiritual traditions. Experience, not belief, is central to Druid spirituality, and so it actually doesn't matter that much to Druidry whether gods are objectively real individual divine beings, aspects or manifestations of some overarching unity, archetypal functions within the human mind, or something else entirely. What matters is that they do certain things, embody certain energies, and appear in certain ways.

Which gods or goddesses a Druid worships is, similarly, a personal matter based on the Druid's personal relation to the primary powers of the cosmos. Common patterns do exist, however. In many traditions, eight deities from Celtic or quasi-Celtic traditions have places around the wheel of the year, embodying the seasonal energies of the year's eight stations.

The deities most commonly invoked in British Druid Revival circles make up one set, with gods and goddesses alternating in a balanced pattern. The descriptions and meanings given below are from twentieth-century Druid sources.

FESTIVAL	DEITY
Samhuinn	Ceridwen, goddess of wisdom and keeper of the sacred cauldron
Alban Arthuan	Hu the Mighty, the great Druid god and master of the solar current
Imbolc	Ana, ancestress of the gods and mistress of the telluric current
Alban Eiler	Coel or Celi, god of the life force and master of animals
Belteinne	Niwalen or Elen, goddess of dawn, dusk, and the old straight track
Alban Heruin	Beli or Belinus, hero-god of fire and the Sun
Lughnasadh	Sul, healing goddess of the Sun and sacred springs
Alban Elued	Esus, chief of tree spirits, who sits in the first fork of the oak

Another sequence, following the same alternation of gods and goddesses, can be drawn readily from Old Irish mythic lore.

FESTIVAL	DEITY
Samhuinn	Morrigan, raven-goddess of war
Alban Arthuan	Dagda Mor, king of the gods
Imbolc	Brighid, goddess of crafts
Alban Eiler	Aenghus Og, god of youth and love
Belteinne	Eriu, goddess of sovereignty
Alban Heruin	Lugh, the divine Sun-hero
Lughnasadh	Tailtiu, goddess of agriculture
Alban Elued	Mannanan, god of the sea

Yet a third set of gods and goddesses, assembled from a wild assortment of sources, appears in the teachings of the Reformed Druids of North America (RDNA). The RDNA is among the least doctrinaire of Druid groups (which is saying something), and its ritual calendar, like nearly everything else, varies wildly from grove to grove. One workable correspondence between gods and holy days, however, is shown below. The alternation here is not between genders, but, broadly speaking, between water and fire.

FESTIVAL	DEITY
Samhuinn	Llyr, god of the sea
Alban Arthuan	Taranis, god of lightning and thunder
Imbolc	Sirona, goddess of rivers
Alban Eiler	Dalon ap Landu, god of groves
Belteinne	Grannos, god of healing springs
Alban Heruin	Belenos, god of the Sun
Lughnasadh	Bracacia, god(dess) of malt and brewing
Alban Elued	Danu, goddess of fertility

One of these gods, Dalon ap Landu, doesn't appear in any earlier source at all, and was probably invented out of whole cloth by the Carleton College undergraduates who concocted the RDNA. What makes this interesting is that Druids in the RDNA tradition have had powerful experiences of Dalon ap Landu's presence. This presents a pretty theological puzzle: Did the god of forests, already called by so many names, simply decide to respond to this one as well? Or did the RDNA in some baffling way call a new god into being? Late night talk around Druid campfires strays into matters such as these.

The Arthurian legends offer another basis for systems of this sort. According to many scholars, they contain a good deal of Celtic Pagan tradition in fragmentary form, and Druids have made much use of them over the years. The knights of Camelot may not be gods, at least as they appear in medieval romances, but they can be assigned places at the Table Round of the year. The fair and loathly ladies of the legends also have a place at the Table, sometimes paired with their knightly lovers, sometimes with knights who share their central characteristics.

FESTIVAL	DEITY
Samhuinn	Sir Kay and Lady Cundrie
Alban Arthuan	Sir Lancelot and Queen Guenivere
Imbolc	Sir Bedivere and the Lady of the Lake
Alban Eiler	Sir Galahad and Lady Dindrane
Belteinne	Sir Tristram and Lady Iseult
Alban Heruin	Sir Gawain and Lady Ragnall
Lughnasadh	Sir Bors and Lady Elaine
Alban Elued	Sir Perceval and Lady Blanchefleur

In this system, King Arthur is the Sun, illuminating the place of each knight in turn, with Lancelot as his alter ego filling his place in the seat of his winter birth and death.

Finally, a similar list can be compiled of saints and angels central to the traditions of Celtic Christianity. While some people in today's Druid scene react heatedly to the idea, the fact remains that there have been plenty of Christian Druids, from the first days of the Revival to the present. There's no good reason to exclude them from the tradition or pretend that Christian saints and angels can't be invoked in a Druid setting. Holy powers from Christian mythology can readily be mapped onto a wheel of the year slightly modified to fit the traditional church calendar, with Easter, Pentecost, and other movable feast days shifting around the year according to Christian custom.

FESTIVAL	SAINT OR ANGEL
All Hallows (1 November)	St. Peter the Apostle
Christmas (25 December)	St. Uriel the Archangel
Candlemas (2 February)	St. Brigid of Kildare
Annunciation (25 March)	St. Gabriel the Archangel
Our Lady's Day (1 May)	St. John the Evangelist
St. John Baptist's Day (24 June)	St. Raphael the Archangel
St. Mary's Day (15 August)	St. Raphael the Archangel
Michaelmas (29 September)	St. Michael the Archangel

It's important to be clear about what is and isn't being said by tabulations of this sort. The common mistake to avoid here is the assumption that things associated with the same thing are equal to each other. The fact that Eriu, Sir Perceval, and St. John the Evangelist are all associated with Belteinne doesn't mean that these three figures are one and the same! Rather, the goddess of sovereignty, the knight of the Grail, and the beloved disciple all correspond to this phase of the yearly cycle, but each expresses that symbolism in a unique way.

These correspondences, however, teach a crucial lesson. The goddess Eriu, for example, has a special link with Belteinne, but is present throughout the year. The same is true of all the gods, goddesses, angels, and saints listed above. And if the knights and ladies of Camelot are seen as personified inner forces, as indeed they should be, they also have a place whatever the season. The same, ultimately, is true of the eight stations of the Sun themselves. Belteinne is a day in the spring, in other words, but it's also a pattern of energy always present in the world.

Thus the wheel of the eightfold year is not just a calendar or a framework for rituals. It's a cosmogram, or diagram of the universe of human experience, traced out in time rather than space. Like other such diagrams used in spiritual traditions—the mandalas of esoteric Buddhism or the Cabalistic Tree of Life—the Wheel of Life, as the pattern of the eightfold year may be called, provides a map of spiritual powers and realms, a way of sorting out the complicated phenomena of inner experience, and a chart for inner journeys from one realm of energy to another (see figure 5). From these possibilities, much of the deeper work of Druid magic unfolds.

FIGURE 5. THE DRUID WHEEL OF LIFE.

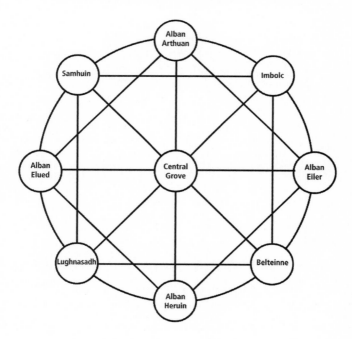

The Wheel of Life, as one might call the eightfold year in its cosmogrammatic form, can be filled out with symbolism as shown in Table 4, with the gods and other powers listed as additional symbolism.

The rich symbolism of the eightfold year and the Wheel of Life cosmogram makes them valuable tools in modern Druidry. At the same

TABLE 5A. SYMBOLISM OF THE DRUID WHEEL OF LIFE				
Festival	Belteinne	Alban Heruin	Lughnasadh	Alban Elued
Planet	Jupiter	Mars	Venus	Mercury
Element	Water	Fire	Air	Spirit
Colors	Sky blue and orange	Red and yellow	Gold and green	Leaf green and blue
Inner place	Holy well	Stone circle	Sacred grove	Standing stone
Emblem	Hirlas (Horn of Mead)	Three rays of light	Turning fiery wheel	Silver branch
Symbol	Flowering hawthorn	Vervain	Ripe grain	Fruit and wine
Energy center	Vital center (two inches below navel)	Solar plexus center (inside upper abdomen, just below breastbone)	Heart center	Throat center (at base of throat)

time, of course, the eightfold wheel of the year is not the only possible Druid calendar, nor the only one currently practiced. Many of the older Druid orders still keep the fourfold wheel of solstices and equinoxes as their ceremonial calendar, while some more recent traditions use a thirteen-month calendar based on the Ogham alphabet. These and others are workable conceptions, and can be used in the same way as the eightfold wheel as a cosmic diagram. As so often in Druid tradition, the Awen moves where it wills, and no one rhythm governs all the movements of its intricate dance.

TABLE 5B. SYMBOLISM OF THE DRUID WHEEL OF LIFE				
Festival	Samhuinn	Alban Athuan	Imbolc	Alban Eiler
Planet	Moon	Sun	Earth	Saturn
Element	Gwyar (essential fluidity)	Nwyfre (essential energy)	Calas (primal substance)	Earth
Colors	Violet and silver	Brilliant white and gold	Earth colors and black	Indigo and red
Inner place	Mountain summit	Starry heavens	Cavern	Interior of sacred mound
Emblem	Cauldron	Crown of light	Circle of eight candles	Dragon
Symbol	Evergreen branches	Mistletoe	Snowdrops	Trefoil
Energy center	Brow center (behind bridge of nose, in center of head)	Crown center (above crown of head)	Earth center (at soles of feet)	Sexual center (at genitals)

Ogham, The Alphabet of the Trees

The ancient Irish alphabet called Ogham (pronounced "OH-am") provides another central theme of modern Druidry—one that operates in harmony with the two powers and the Wheel of Life. The oldest surviving Ogham writings are carved stones from the third century C.E., but references in Irish legend place its use well back in Pagan times and suggest that it was normally written on wood, which rarely survives long in the wet Irish climate. According to medieval Irish Ogham treatises preserved in the *Book of Ballymote*, some of the letters stand for sounds that don't exist in the Irish language at all, which only

83

deepens the mystery. Was Ogham patterned on some other writing system, such as Latin or Greek? Or did it originate, as some modern scholars suggest, among speakers of a pre-Celtic Irish language in megalithic times? Nobody knows.

Whatever its origins, Ogham is an oddity among the world's alphabets. Its inscriptions read from bottom to top in vertical lines. Its twenty basic letters, called *fews* (Old Irish *feadha*), consist of tally marks in groups of one to five—first to the right of a line, then to the left, then diagonally across, then straight across (see figure 6). Each group of five fews is called an *aicme*. The order of the original twenty letters—BLNVS HDTCQ MGNgZR AOUEI—has no relation to any other known alphabet's order, though the first aicme seems to spell a familiar name: BeLeNVS or Belenos, the Celtic god of light. (It's probably not accidental that, in Christian times, the first aicme was reordered BLVSN.) A group of five *forfedha* ("extra letters") with various shapes and meanings was added to the alphabet sometime later. Weird as this may sound, the simple shapes of its letters make Ogham an excellent alphabet for writing on wood or stone with simple tools.

The great value of Ogham to the modern Druid lies in its wealth of symbolism. Each few is linked in old Irish lore with a tree (not all are "trees" in the modern English sense of the word). Each few also has a bird, a color, an agricultural tool, and an art associated with it. The old Irish word for each of these begins with the few, and a set of animals can readily be added along the same lines. Other symbolism has been added in recent centuries by Druids working with the alphabet. In this way, Ogham serves as a magical alphabet in divination, ritual, and many other forms of Druid practice. Some purists have denounced the additions, but the Irish Ogham texts themselves point out that students should rework the Ogham symbols to suit their own experience and needs.

FIGURE 6. THE OGHAM ALPHABET.

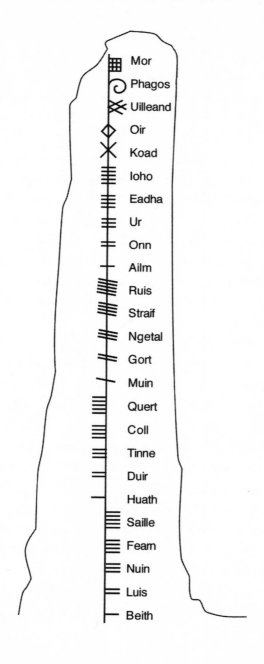

Mor
Phagos
Uilleand
Oir
Koad
Ioho
Eadha
Ur
Onn
Ailm
Ruis
Straif
Ngetal
Gort
Muin
Quert
Coll
Tinne
Duir
Huath
Saille
Fearn
Nuin
Luis
Beith

One detail of Ogham symbolism deserves discussion. In the 1940s, the English poet Robert Graves came across Ogham while studying Celtic mythology. His source, an eighteenth-century book of Irish history, listed thirteen of the fews, and Graves noticed that the trees linked to those fews formed a seasonal sequence. With this as his starting point, he worked out a thirteen-month tree calendar with twenty-eight days to each month, and one extra day at the winter solstice assigned to the sacred mistletoe.

Graves believed that he'd stumbled across one of the secrets of the ancient Druids. No trace of his calendar occurs anywhere in folklore or historical sources, however, and the very meager scraps of ancient Celtic calendar lore that have survived suggest that the Pagan Celts used quite different calendars. Yet the tree calendar works, and works very well, as a Druid time-keeping tool and as a way of expanding the symbolism of Ogham. Astrologers have even made it the basis of a new thirteen-sign zodiac that works as well as the traditional system and has been adopted by some modern Druids.

In other words, the equation between time and the timeless explored earlier in this chapter works in both directions. Just as the eight holy days can rise out of the seasonal cycle to become ever-present patterns of meaning, the Ogham fews can descend into time's circle and become markers for the changes of the turning year. None of this proves that the ancient Druids used Ogham in this way, of course. What it does prove is that modern Druids use Ogham in this way, and find it worthwhile.

Below is an outline of the basic symbolism of the Ogham alphabet. The animals assigned to each few are original additions based on traditional lore; the calendar attributions and one bird (the crane for Coll) come from Robert Graves; the elements, the paths on the Wheel of Life, and the divinatory meanings are modern. The rest of the symbolism comes from the old Irish Ogham texts. The word following the name and pronunciation of each few is the name's literal meaning, which (despite an assumption common in many modern circles) is not always the name of its associated tree! The word Oghams farther down are traditional poetic metaphors, supposedly created by the famous judge Morann mac Main, the warrior Cu Chulainn, and the god Aenghus Og, under his title Mac ind Óic (Son of Og). The remainder is self-explanatory.

Much of the old lore is confusing, and not all of it has been translated satisfactorily by scholars. Comments have been added in parentheses where this seemed helpful. For the rest, Druids who have studied botany know that "spp." after a single-word scientific name of a plant means all the species belonging to that genus; thus "*Betula* spp." means all the species of the *Betula*, or birch, genus.

BEITH (pronounced "BEH"): Being

A few of beginnings and purification.

Tree: Birch (*Betula* spp.)

Sound value: B

Word Ogham of Morann mac Main: "Faded trunk and fair hair."

Word Ogham of Cú Chulainn: "Browed beauty, that is, worthy of pursuit."

Word Ogham of Mac ind Óic: "Most silver of skin."

Number Ogham of Bricriu: 1

Color Ogham: *Ban*, white (a dull white like birchbark; pure white is assigned to Ioho)

Bird Ogham: *Besan*, pheasant

Animal Ogham: *Bo*, cow

Tool Ogham: *Biaill*, axe

Art Ogham: *Bethumnacht*, livelihood (the basic skills of living)

Elemental attribution: Spirit of Air

Path of the Wheel of Life: Alban Arthuan to Imbolc

Calendar attribution: First month, 24 December–20 January

Upright divinatory meanings: Beginnings, new possibilities, potentials; renewal and rebirth; a favorable sign in most matters, though there may be discomforts involved.

Reversed divinatory meanings: Blind alleys, wasted effort; creative blockages; remaining fixated on the past or on things that have been lost; longing for the impossible.

LUIS (pronounced "LWEESH"): Herb

A few of protection, discernment, and inner clarity.

Tree: Rowan (*Sorbus* spp.)

Sound value: L

Word Oghamof Morann mac Main: "Delight of the eye"

Word Ogham of Cú Chulainn: "Strength of cattle, that is, the elm"

Word Ogham of Mac ind Óic: "Friend of cattle"

Number Ogham of Bricriu: 2

Color Ogham: *Liath*, gray

Bird Ogham: *Lachu*, duck

Animal Ogham: *Luchog*, mouse

Tool Ogham: *Loman*, rope

Art Ogham: *Luamnacht*, sailing

Elemental attribution: Fire of Air

Path of the Wheel of Life: Alban Arthuan to Alban Eiler

Calendar attribution: Second month, 21 January–17 February; includes Imbolc, 2 February

Upright divinatory meanings: Clarity and attention to details; concentration on the task at hand; purification; danger avoided or overcome; a choice between two paths.

Reversed divinatory meanings: Confusion, uncertainty,

deception, delusion; lack of defenses; something is not what it appears to be; someone may be misleading you, or you may be misleading yourself.

NUIN (pronounced "NOO-un"): Letter

A few of connection, communication, and magic.

Tree: Ash (*Fraxinus* spp.)

Sound value: N

Word Ogham of Morann mac Main: "Checking of peace"

Word Ogham of Cú Chulainn: "Flight of beauty, that is, a weaver's beam"

Word Ogham of Mac ind Óic: "Flight of women"

Number Ogham of Bricriu: 3

Color Ogham: *Necht*, clear

Bird Ogham: *Naescu*, snipe

Animal Ogham: *Naddair*, adder

Tool Ogham: *Nasc*, ring

Art Ogham: *Notaireacht*, notary work (more generally, any form of recordkeeping)

Elemental attribution: Air of Air

Path of the Wheel of Life: Imbolc to Alban Eiler

Calendar attribution: Third month, 18 February–17 March

Upright divinatory meanings: Connection and transformation; relationship to a larger context; you are part of a wider world; events may be more important than they appear.

Reversed divinatory meanings: Isolation and self-containment; you are separated from what is going on, for good or ill; boredom and lack of initiative may blind you to the possibilities of the present.

FEARN (pronounced "FAIR-n"): Alder

A few of oracular guidance, protection, and transitions from realm to realm.

Tree: Alder (*Alnus* spp.)

Sound value: V

Word Ogham of Morann mac Main: "Defense of warrior bands"

Word Ogham of Cú Chulainn: "Protection of the heart, that is, a shield"

Word Ogham of Mac ind Óic: "Guarding of milk"

Number Ogham of Bricriu: 4

Color Ogham: *Flann*, red

Bird Ogham: *Faelinn*, seagull

Animal Ogham: *Faol*, wolf

Tool Ogham: *Fidba*, hedge-bill (a hooked iron blade with a handle used to trim hedges)

Art Ogham: *Filidecht*, bardic poetry

Elemental attribution: Water of Air

Path of the Wheel of Life: Imbolc to Belteinne

Calendar attribution: Fourth month, 18 March–14 April; includes Alban Eiler, 20 March

Upright divinatory meanings: Protection and guidance, a bridge over deep waters; steadfastness; good advice received from others or your own inner depths; spiritual guidance and insight, the presence of the gods; an unexpected way past a difficulty.

Reversed divinatory meanings: Willful blindness, refusal to listen to advice; arrogance and lack of insight; you are out of your depth.

SAILLE (pronounced "SAHL-yuh"): Willow

A few of grace, fluidity, receptivity, and response.

Tree: Willow (*Salix* spp.)

Sound value: S

Word Ogham of Morann mac Main: "Hue of the lifeless"

Word Ogham of Cú Chulainn: "Beginning of loss, that is, willow"

Word Ogham of Mac ind Óic: "Strength of bees"

Number Ogham of Bricriu: 5

Color Ogham: *Sodath*, fine-colored (many modern Druids use primrose yellow, *serind* in Irish; for Saille's color)

Bird Ogham: *Segh*, hawk

Animal Ogham: *Sionnach*, fox

Tool Ogham: *Srathar*, pack saddle

Art Ogham: *Sairsi*, handicrafts (all simple craft work using hand tools)

Elemental attribution: Earth of Air

Path of the Wheel of Life: Alban Eiler to Belteinne

Calendar attribution: Fifth month, 15 April–12 May; includes Belteinne, 1 May

Upright divinatory meanings: Moving with the flow of events; intuition, dreaming, the unconscious; letting go of fixed ideas and responding to the moment; faith in a personal vision.

Reversed divinatory meanings: Inability to move with the situation due to rigidity; Confusion or lack of awareness; unforeseen dangers; a difficult time that will pass eventually.

HUATH (pronounced "OO-ah"): Terrible

A few of patience, restriction, and desire not yet fulfilled.

Tree: Hawthorn (*Crataegus* spp.)

Sound value: H

Word Ogham of Morann mac Main: "Pack of wolves"

Word Ogham of Cú Chulainn: "Difficult night, that is, hawthorn"

Word Ogham of Mac ind Óic: "Whitening of face"

Number of Bricriu: 6

Color Ogham: *h-Uath*, terrible (many modern Druids use deep violet for Huath's color)

Bird Ogham: *Hadaig*, raven

Animal Ogham: *h-Aodh*, sheep

Tool Ogham: *Huartan*, meaning uncertain (possibly a salt-box?)

Art Ogham: *h-Airchetul*, trisyllabic poetry (poetry in a meter of three syllables to a line)

Elemental attribution: Spirit of Fire

Path of the Wheel of Life: Alban Eiler to Alban Heruin

Calendar attribution: Sixth month, 13 May–9 June

Upright divinatory meanings: Patience, reserve, retreat; a time of waiting and planning rather than action; obstacles and entanglements that can be overcome; success after a delay; temporary obstacles.

Reversed divinatory meanings: Inappropriate action, rushing ahead when patience and planning are called for; a risk of failure; you need to stop and reconsider.

DUIR (pronounced "DOO-er"): Oak

A few of power, protection, and change for the better.

Tree: Oak (*Quercus* spp.)

Sound value: D

Word Ogham of Morann mac Main: "Highest of bushes"

Word Ogham of Cú Chulainn: "Kneeling work, that is, bright and shining work"

Word Ogham of Mac ind Óic: "Craft work"

Number of Bricriu: 7

Color Ogham: *Dubh*, black

Bird Ogham: *Droen*, wren

Animal Ogham: *Damh*, stag

Too Ogham: *Dabach*, cask (that is, a small barrel or container)

Art Ogham: *Druidheacht*, Druidry

Elemental attribution: Fire of Fire

Path of the Wheel of Life: Belteinne to Alban Heruin

Calendar attribution: Seventh month, 10 June–7 July; includes Alban Heruin, 21 June

Upright divinatory meanings: Success, forward movement, attainment; power and energy; you have all you need to accomplish your goals; a door opens in the outer world.

Reversed divinatory meanings: Help from those in positions of power or authority; success despite inadequate means; borrowed strength; a gift.

TINNE (pronounced "CHIN-yuh"): Iron

A few of courage, conflict, and opposition.

Tree: Holly (*Ilex* spp.)

Sound value: T

Word Ogham of Morann mac Main: "Third of a wheel"

Word Ogham of Cú Chulainn: "Third of weapons, that is, an iron bar"

Word Ogham of Mac ind Óic: "Fires of coal"

Number Ogham of Bricriu: 8

Color Ogham: *Temen*, dark gray

Bird Ogham: *Truiteoc*, starling

Animal Ogham : *Torc,* boar

Tool Ogham: *Tal,* adze

Art Ogham: *Tornoracht,* lathe work

Elemental attribution: Air of Fire

Path of the Wheel of Life: Belteinne to Lughnasadh

Calendar attribution: Eighth month, 8 July–4 August; includes Lughnasadh, 1 August

Upright divinatory meanings: Conflict, challenge, struggle against opposing forces; victory against the odds; a change of fortune; decisive action is favored.

Reversed divinatory meanings: Inadequate strength or skill, the possibility of defeat; lack of direction and balance; you need to build your strength and understand the nature of the opposition.

COLL (pronounced "CULL"): Hazel

A few of knowledge, creativity, and inspiration.

Tree: Hazel (*Corylus* spp.)

Sound value: C

Word Ogham of Morann mac Main: "Fairest of trees"

Word Ogham of Cú Chulainn: "Sweetest of woods, that is, a nut"

Word Ogham of Mac ind Óic: "Friend of cracking"

Number Ogham of Bricriu: 9

Color Ogham: *Cronn*, brown

Bird Ogham: *Corr*, crane

Animal Ogham: *Cat*, cat

Tool Ogham: *Carr*, wagon

Art Ogham: *Cruitireacht*, harping

Elemental attribution: Water of Fire

Path of the Wheel of Life: Alban Heruin to Lughnasadh

Calendar attribution: Ninth month, 5 August–1 September

Upright divinatory meanings: Knowledge, intelligence, talent; transformation and flexibility; the beginning of a new stage in life; communication and teaching, new information.

Reversed divinatory meanings: Creative and intellectual blockages, lack of insight; often, fear of failure, leading to a retreat to familiar ground.

QUERT (pronounced "KWEIRT"): Apple

A few of delight, celebration, and choice.

Tree: Apple (*Pyrus malus, Pyrus sylvestris*)

Sound value: Q

Word Ogham of Morann mac Main: "Shelter of a hind, that is, shelter of lunatics"

Word Ogham of Cú Chulainn: "Excellent emblem, that is, protection"

Word Ogham of Mac ind Óic: "Force of a man"

Number Ogham of Bricriu: 10

Color Ogham: *Quiar*, mouse-brown

Bird Ogham : *Querc*, hen

Animal Ogham: *Cu*, dog

Tool Ogham: *Cual*, stick

Art Ogham: *Quislenacht*, flute-playing

Elemental attribution: Earth of Fire

Path of the Wheel of Life: Alban Heruin to Alban Elued

Calendar attribution: (no calendar attribution)

Upright divinatory meanings: Happiness, healing, and recovery; awakenings and new experiences; an unexpected gift; the rewards of success; an opportunity to live more fully.

Reversed divinatory meanings: An unavoidable choice among alternatives; mixed gain and loss; a temptation to scattered effort or procrastination that must be overcome.

MUIN (pronounced "MUHN"): Back

A few of insight and inspiration.

Tree: Grapevine (*Vitis vinifera*)

Sound value: M

Word Ogham of Morann mac Main: "Strongest of effort"

Word Ogham of Cú Chulainn: "Wolfpack with spears, that is, three vines"

Word Ogham of Mac ind Óic: "Condition of slaughter"

Number Ogham of Bricriu: 11

Color Ogham: *Mbracht*, variegated (many modern Druids use plaid for Muin's color)

Bird Ogham: *Mintan*, titmouse (a small bird related to the chickadee)

Animal Ogham: *Muc*, sow

Tool Ogham: *Machad* (meaning uncertain—perhaps a milking bucket)

Art Ogham: *Milaideacht*, soldiering

Elemental attribution: Spirit of Water

Path of the Wheel of Life: Lughnasadh to Alban Elued

Calendar attribution: Tenth month, 2 September–29 September; includes Alban Elued, 22 September

Upright divinatory meanings: Inspiration and prophecy; community; the influence of spiritual factors on the situation; unexpected truths; freedom from limits and restrictions.

Reversed divinatory meanings: Burdens, difficulties; a need to relax and unwind; you have been trying too hard in unproductive ways.

GORT (pronounced "GORT"): Field

A few of tenacious purpose and indirect progress.

Tree: Ivy (*Helix hedera*)

Sound value: G

Word Ogham of Morann mac Main: "Sweeter than grasses, that is, a field of grain"

Word Ogham of Cú Chulainn: "Pleasing oil, that is, grain"

Word Ogham of Mac ind Óic: "Size of a warrior"

Number Ogham of Bricriu: 12

Color Ogham: *Gorm*, sky blue

Bird Ogham: *Geis*, mute swan

Animal Ogham: *Grainneog*, hedgehog

Tool Ogham: *Gat*, withe (a flexible length of willow twig, used in place of cordage)

Art Ogham: *Gaibneacht*, blacksmithying

Elemental attribution: Fire of Water

Path of the Wheel of Life: Lughnasadh to Samhuinn

Calendar attribution: Eleventh month, 30 September–27 October

Upright divinatory meanings: Slow and indirect progress, movement by roundabout paths; purpose, determination, self-control; a difficult but viable path.

Reversed divinatory meanings: Entanglement in circumstances, or in your own egotism; think twice about what you are doing, and why.

NGETAL (pronounced "NYEH-tal"): Wound

A few of transformation and healing.

Tree: Reed (*Phragmites* spp.) or Broom (*Genista* spp.)

Sound value: Ng

Word of Morann mac Main: "Strength of a healer"

Word of Cú Chulainn: "Beginning of heroic deeds, that is, healing"

Word Ogham of Mac ind Óic: "Robe of healers"

Number Ogham of Bricriu: 13

Color Ogham: *nGlas*, grass green

Bird Ogham: *nGeigh*, goose

Animal Ogham: *nGearr*, hare

Tool Ogham: *nGend*, wedge

Art Ogham: *nGibae*, sculpting

Elemental attribution: Air of Water

Path of the Wheel of Life: Alban Elued to Lughnasadh

Calendar attribution: Twelfth month, 28 Oct–24 November; Includes Samhuinn, 1 November

Upright divinatory meanings: Swiftness, sudden movement, instability; healing, change for the better; a message or an opportunity; you are not yet finished with your work.

Reversed divinatory meanings: Stuck in a rut, inability to act; a need for healing; patience; you are not yet ready to begin.

STRAIF (pronounced "STRAHF"): Sulphur

A few of necessity and inevitable change.

Tree: Blackthorn (*Prunus spinosa*)

Sound value: Z

Word Ogham of Morann mac Main: "Careful effort, strongest of red"

Word Ogham of Cú Chulainn: "Mist of an arrow, that is, smoke rising from a fire"

Word Ogham of Mac ind Óic: "Increasing of secrets"

Number Ogham of Bricriu: 14

Color Ogham: *Sorcha*, bright (many modern Druids use silver for Straif's color)

Bird Ogham: *Stmolach*, thrush

Animal Ogham: *Srianach*, badger

Tool Ogham: *Sust*, flail (two sticks hinged together, used to separate grain from husks)

Art Ogham: *Sreghuindeacht*, deer hunting

Elemental attribution: Water of Water

Path of the Wheel of Life: Alban Elued to Alban Arthuan

Calendar attribution: (no calendar attribution)

Upright divinatory meanings: Necessity and inevitability; the hard realities of life; something that cannot be changed; the results of one's own actions; the influence of fate.

Reversed divinatory meanings: Pain, difficulty, retribution, unavoidable suffering; a difficult path that must be taken; every choice leads to unhappiness.

RUIS (pronounced "RWEESH"): Reddening

A few of resolutions, fulfillments, and endings.

Tree: Elder (*Sambucus* spp.)

Sound value: R

Word Ogham of Morann mac Main: "Most intense of blushes"

Word Ogham of Cú Chulainn: "Harsh anger, that is, punishment"

Word Ogham of Mac ind Óic: ""Redness of faces"

Number Ogham of Bricriu: 15

Color Ogham: *Ruadh*, dark red

Bird Ogham: *Rocnat*, rook (a variety of crow)

Animal Ogham: *Ron*, seal

Tool Ogham: *Rusc*, basket

Art Ogham: *Ronnaireacht*, pharmacy (preparing any healing substance)

Elemental attribution: Earth of Water

Path of the Wheel of Life: Samhuinn to Alban Arthuan

Calendar attribution: Thirteenth month, 25 November–22 December

Upright divinatory meanings: Healing, resolution, completion; transitions from one state of being to another; difficulties permanently overcome; an omen of success in most things. Reversed divinatory meanings: Endings and departures; the need to release things that are past their time; facing up to facts; embarrassment and shame; illness and disability.

AILM (pronounced "AHL-m"): Elm

A few of vision, understanding, and eminence.

Tree: Fir (*Abies* spp.) or Elm (*Ulmus* spp.)

Sound value: A

Word Ogham of Morann mac Main: "Loudest of groanings"

Word Ogham of Cú Chulainn: "Beginning of weaver's beams, that is, A"

Word Ogham of Mac ind Óic: "Beginning of answers"

Number Ogham of Bricriu: 16

Color Ogham: *Alad*, piebald (that is, spotted black and white)

Bird Ogham: *Aidhircleog*, lapwing

Animal Ogham: *Art*, bear

Tool Ogham: *Arathar*, plow

Art Ogham: *Airigeacht*, leadership

Elemental attribution: Spirit of Earth

Path of the Wheel of Life: Alban Arthuan to Central Grove

Calendar attribution: Alban Arthuan, 23 December

Upright divinatory meanings: Insight, transformation,

expanded awareness; change for the better; the ability to see things in perspective; peak experiences, dreams and visions.

Reversed divinatory meanings: Lack of perspective, ignorance of the broader picture; unrealistic ideas; you need to step back and assess the situation more carefully.

ONN (pronounced "UHN"): Gorse

A few of attraction, combination, and possibility.

Tree: Gorse (*Ulex* spp.)

Sound value: O

Word Ogham of Morann mac Main: "Helper of horses, wheels of the chariot"

Word Ogham of Cú Chulainn: "Strength of a warrior band, that is, fierceness"

Word Ogham of Mac ind Óic: "Gentlest of work"

Number Ogham of Bricriu: 17

Color Ogham: *Odhar*, dun (that is, light yellowish brown)

Bird Ogham: *Odoroscrach*, cormorant

Animal Ogham: *Os*, deer

Tool Ogham: *Ord*, hammer

Art Ogham: *Ogmoracht*, harvesting

Elemental attribution: Fire of Earth

Path of the Wheel of Life: Alban Eiler to Central Grove

Calendar attribution: Alban Eiler, 20 March

Upright divinatory meanings: Gathering together, combination of forces; energy, life, vigor, sexuality, and attraction; opportunities, though not without potential problems.

Reversed divinatory meanings: Difficulties and delays, overconfidence; desires out of touch with the realities of the situation; uncoordinated efforts; problems, but with potential benefits.

UR (pronounced "OOR"): Earth

A few of power, creation, death, and rebirth.

Tree: Heather (*Erica* spp., *Calluna* spp.)

Sound value: U

Word Ogham of Morann mac Main: "Terrible tribe in cold dwellings"

Word Ogham of Cú Chulainn: "Completion of lifelessness, that is, the grave"

Word Ogham of Mac ind Óic: "Growing of plants"

Number Ogham of Bricriu: 18

Color Ogham: *Usgdha*, resin-colored (a reddish amber color)

Bird Ogham: *Uiseoc*, lark

Animal Ogham: *Urcuil*, cricket

Tool Ogham: *Usca*, heather brush (heather was used to make brushes in Ireland)

Art Ogham: *Umaideacht*, brasswork

Elemental attribution: Air of Earth

Path of the Wheel of Life: Alban Heruin to Central Grove

Calendar attribution: Alban Heruin, 21 June

Upright divinatory meanings: Spiritual power, the forces of nature; fulfillment; the coming of new life; a door opens in the inner world; passion, power, and magic; creation.

Reversed divinatory meanings: Weakness, dispersed energies, loss of contact with nature and the spirit; dreamy neglect of realities; a need for recuperation and reconnection.

EADHA (pronounced "EH-yuh"): Aspen

A few of perseverance, courage, and hard work.

Tree: Aspen (*Populus tremula, Populus tremuloides*)

Sound value: E

Word Ogham of Morann mac Main: "Distinguished man or wood"

Word Ogham of Cú Chulainn: "Kinsman to the birch, that is, aspen"

Word Ogham of Mac ind Óic: "Another name for a friend"

Number Ogham of Bricriu: 19

Color Ogham: *Erc*, rust red

Bird Ogham: *Ela*, whistling swan

Animal Ogham: *Each*, horse

Tool Ogham: *Epit*, billhook (a hooked blade on a handle, used to trim tree limbs)

Art Ogham: *Enaireacht*, bird hunting

Elemental attribution: Water of Earth

Path of the Wheel of Life: Alban Elued to Central Grove

Calendar attribution: Alban Elued, 22 September

Upright divinatory meanings: Courage and tenacity in the face of opposition; a struggle in which victory is possible but not certain; a quest for inner strength.

Reversed divinatory meanings: Declining strength; compromise and negotiation; prudence; if you continue in your present path, the results will not be good.

IOHO (pronounced "EE-yoh"): Yew

A few of enduring realities and legacies.

Tree: Yew (*Taxus* spp.)

Sound value: I, Y

Word Ogham of Morann mac Main: "Oldest of woods"

Word Ogham of Cú Chulainn: "Color of a sick man, that is, a people (or an age)"

Word Ogham of Mac ind Óic: "Abuse for an ancestor, or pleasing consent"

Number of Bricriu: 20

Color Ogham: *Irfind*, very white (the color of newly fallen snow)

Bird Ogham: *Illait*, eaglet

Animal Ogham: *Ialtog*, bat

Tool Ogham: *Indeoin*, anvil

Art Ogham: *Iascaireacht*, fishing

Elemental attribution: Earth of Earth

Path of the Wheel of Life: Samhuinn to Imbolc

Calendar attribution: Alban Arthuan, 23 December

Upright divinatory meanings: Enduring realities, that which remains unchanged; old age, legacies from the past; the consequences of present actions.

Reversed divinatory meanings: Stagnation and immobility; things lingering past their time; the past as a burden not yet overcome.

The five forfedha, or extra letters, have a much less fully developed symbolism in the Irish sources, and their sound values range all over the phonetic map. The specific values given below are the most useful for Druids who speak modern English. The colors were introduced by Colin Murray of the Golden Section Order, who also introduced the practice of Ogham divination to modern Druidry. The calendar attributions are original, though based on older symbolism.

KOAD (pronounced "KO-ud"): Grove

A few of central balance and infinite possibility.

Tree: A grove containing all trees.

Alternate name: Ebhadh (pronounced "EV-uth"); poplar

Alternate tree: Elecampane (*Inula helenium*)

Sound value: Ch

Color Ogham: Many shades of green

Elemental attribution: Spirit of Spirit

Path of the Wheel of Life: Central Grove of the Wheel

Calendar attribution: The whole cycle of the year

Upright divinatory meanings: Complexity, the presence of many factors; possibility; capacity for freedom.

Reversed divinatory meanings: Confusion and bewilderment; there may be too many factors at work to allow successful prediction.

OIR (pronounced "OR"): Spindle

A few of fate, sudden change, and the unexpected.

Tree: Spindle (*Euonymus* spp.)

Alternate name: Tharan (pronounced "TAR-ann"); thunder

Sound value: Th

Color Ogham: Brilliant white (like lightning)

Elemental attribution: Fire of Spirit

Path of the Wheel of Life: Imbolc to Central Grove

Calendar attribution: Imbolc, 2 February

Upright divinatory meanings: Radical transformation, the flash of the lightning bolt; sudden change, illumination or destruction, set in motion by forces outside the situation.

Reversed divinatory meanings: Patience and preparation;

the path before you is slow and cannot be hurried; wait for outside forces to act.

UILLEAND (pronounced "ULL-enth"): Honeysuckle

A few of secrets hidden and revealed.

Tree: Honeysuckle (*Lonicera* spp.)

Alternate name: Peith (pronounced "PETH"); water elder

Alternate tree: Water elder (*Viburnum* spp.)

Sound value: P

Color Ogham: Yellow-white

Elemental attribution: Air of Spirit

Path of the Wheel of Life: Belteinne to Central Grove

Calendar attribution: Belteinne, 1 May

Upright divinatory meanings: Secrets and revelations, the influence of the subtle and seemingly insignificant; small causes with large effects; insight into the nature of the situation.

Reversed divinatory meanings: The situation is not what it seems; hidden factors are a source of complications; the information you need is not available to you.

PHAGOS (pronounced "FAH-gus"): Beech

A few of learning and guidance.

Tree: Beech (*Fagus* spp.)

Alternate name: Iphin (pronounced IFF-in); vineyard

Alternate tree: Gooseberry (*Ribes grossularia*)

Sound value: Ph or F

Color Ogham: Orange-brown

Elemental attribution: Water of Spirit

Path of the Wheel of Life: Lughnasadh to Central Grove

Calendar attribution: Lughnasadh, 1 August

Upright divinatory meanings: Knowledge and lore, learning, study, education; the wisdom of the past as a guide to the present and future.

Reversed divinatory meanings: Ignorance; lack of attention to existing knowledge; failure to learn from experience.

MÓR (pronounced "MOHR"): Sea

A few of beginnings, endings, and the influence of outside forces.

Meaning: Sea

Alternate name: Emancoll (pronounced "EM-un-coll"); coll crossed

Alternate tree: Witch hazel (*Hamamelis* spp.)

Sound value: X

Color Ogham: Blue-green

Elemental attribution: Earth of spirit

Path of the Wheel of Life: Samhuinn to Central Grove

Calendar attribution: Samhuinn, 1 November

Upright divinatory meanings: Unexpected change, the arrival of a new influence; the effects of destiny.

Reversed divinatory meanings: Sudden endings and disruptions, the dissolution of the familiar.

All this symbolism may seem daunting at first, but it provides a framework of images and ideas that can be put to many uses. Spiritual traditions around the world have long used writing systems as potent tools for their work of inner transformation. From Cabalists working with the subtle meanings of the Hebrew alphabet to Japanese mystics using *kotodama* (the secret lore of Japan's syllabic writing system) to unfold the hidden dimensions of Shinto, the basic approach remains the same.

In the Druid tradition, Ogham fills the same function and can be used in nearly all the same ways. Most of its applications belong to levels of Druid practice more advanced than those this book covers, but a solid grounding in Ogham and its symbolism builds the necessary foundation for these levels of work. Learning the Ogham fews by sight is essential, and getting a general sense of each few's symbolic meanings only slightly less so. Regular meditation on the fews and their symbols, using the approach to meditation covered in chapter 6 opens the way to many of the treasures of the Druid path.

6

THE THIRD
TRIAD

THE THIRD TRIAD COVERED
here circles back to the beginning. It consists of three
great mythic figures, each linked to the others, but each
central to a myth of his own—the wizard-bard Taliesin;
Arthur, the once and future king; and Merlin the Mage,
the great archetype of the Druid. Together, they offer a
glimpse into the deeper reaches of the Druid tradition.

Taliesin, the Wizard Bard

Taliesin is the archetypal bard of British myth. Less famous than
Arthur or Merlin, he has, nonetheless, been important in Druid tradition since the beginning of the Revival, and before then had a central
place in the Welsh Druid renaissance of the Middle Ages. The youngest
of the triad according to British folk tradition, Taliesin comes first here
because his tale defines the core pattern of Druid initiation.

The wise sorceress Ceridwen had two children: a daughter named Creirwy, the loveliest of girls, and a son named Afagddu, the ugliest of boys. To compensate her son for his ugliness, Ceridwen set out to make him the wisest person alive. She brewed a potion of wisdom for him in a great cauldron, gathering herbs by the seasons and the movements of the stars. Gwion, a poor orphan boy from a nearby village, was set to stir the potion, which needed a year and a day to brew. When the year and a day was up, three drops of the boiling brew flew out of the cauldron and landed on Gwion's thumb.

The boy put his thumb in his mouth to stanch the pain of the burn and, in that instant, all the virtue of the potion passed into him. He knew everything there was to know—including the fact that Ceridwen would kill him once she learned what had happened. Calling on his newfound magical wisdom, he changed himself into a fleet hare and sped away.

The sorceress soon arrived and, realizing what had happened, changed herself into a greyhound and went in pursuit. Before the hound could catch him, Gwion ran to the water's edge, turned himself into a fish, and swam away. Ceridwen could not be foiled so easily; she turned herself into an otter and darted after him. Before she could catch him, he leapt into the air, turned himself into a bird, and flew off. A moment later, Ceridwen became a hawk and gave chase. With no other hiding place in sight, Gwion dove toward a pile of grain on a threshing floor and turned himself into a single grain of wheat. Ceridwen at once became a hen, leapt on the grain of wheat, and swallowed it.

A short time later, to her dismay, she found that she was pregnant. Knowing the child was Gwion, she planned to put him to death when he was born, but the infant was so beautiful she couldn't bear to do so. Instead, she tied him in a leather sack, tossed him into the river Dyfi, and let the waters take him to his destiny.

Downstream on the Dyfi lay a weir owned by Gwyddno Garanhir, the king of that country. Gwyddno had a son named Elffin who had been plagued by misfortune since his birth. In the hope that some good fortune might come his way, the king set Elffin to watch the weir for a year, with the right to take anything that washed up there. Nothing of value appeared until the day that the leather sack carrying Gwion drifted into the weir. Elffin opened the sack, wondering what might lie within, and found a beautiful newborn boy with light shining about his head. "Behold a shining brow (*tal iesin*)," he said aloud, and was astonished when the baby replied, "Taliesin be it," accepting the words as his name. Elffin took the infant to his wife, who nursed the child and cared for it as though he were her own. From that moment on, all Elffin's misfortune changed to good and the child Taliesin became his household bard.

It happened some time later that Maelgwn, the high king of North Wales, was holding court at Degannwy, surrounded by his nobles and warriors and a great crowd of fawning bards who flattered him constantly. Elffin went there and, in the midst of the flattery, commented to another guest that, while only a king could compete with a king, he had a wife as faithful as any lady in the kingdom, and a bard wiser than all the bards of the king.

This was repeated to King Maelgwn, who flung Elffin into prison until he could prove the truth of his boasts. Maelgwn sent his son, Rhun, to seduce Elffin's wife by trickery, but, thanks to Taliesin's advice, the trick was turned back on Rhun. Then Taliesin went to Degannwy, entered the castle unnoticed, sat in a corner, and cast a spell on all King Maelgwn's bards so that they could do nothing but play "blerwm blerwm" on their lips with their fingers. The king, incensed, had one of his squires hit Heinin, the chief bard, over the head with a broom. This broke the spell and Heinin begged the king's forgiveness, explaining that a spirit in the form of a child had cast an enchantment on him.

King Maelgwn then sent for Taliesin and asked who he was and whence he came. The child-bard replied in a brilliant cascade of poetic riddles that left the king's bards utterly baffled, then asked the king to free Elffin. When Maelgwyn refused, Taliesin summoned a wind so fierce that it terrified the king into sending for his prisoner. Taliesin then conjured the shackles off Elffin with another incantation, and bard and prince went home in triumph.

Like most of the old British legends, the tale of Taliesin wanders freely across the misty borders between myth and history. Taliesin himself was a bard of the late sixth century and the author of some of the oldest classical Welsh poetry. King Maelgwn of North Wales was equally historical, and the early Welsh chronicler Nennius mentions his fawning bards. King Gwyddno Garanhir and his son Elffin are more problematic, since the old chronicles claim that Gwyddno's kingdom was west of the present shores of central Wales, in territory since submerged by the Irish Sea. Modern scientists admit the eastern half of the Irish Sea was dry land in geologically recent times, but date its drowning no later than 1800 B.C.E.—though modern scientists have been wrong before. As for Ceridwen and the more magical dimensions of the legend, it's hard to find an obvious place for these outside of the timeless realm of wonder tale.

The tale of Taliesin has much to offer modern Druids, although many of its lessons can only be unfolded through meditation. Clearly, it has much to say about the importance of the bardic arts of poetry and music in Druidry, both in ancient times and today. Yet the events that changed the ignorant child Gwion into the wizard-poet Taliesin also echo traditions of mystical initiation found all over the world, in which aspirants pass through a transformative process that teaches them to awaken the subtle powers hidden within them.

Gwion first spends a year and a day patiently stirring the cauldron at Ceridwen's bidding. Next, he passes through three transformative journeys that echo the pattern of the three Druid elements: the hare on the earth for calas, the fish in running water for gwyar, and the bird in open air for nwyfre. Then, drawing inward into the core of himself, he becomes a seed and spends nine months in the womb of Ceridwen. Finally, reborn and nameless, he passes through the last ordeal of the waters and awakens into power and vision, with the radiance of the Moon center shining at his brow and all knowledge at his command.

This sequence can be found in the mystical traditions of many places and times, and it also maps closely onto the customs of traditional bardic schools in Ireland and the Scottish highlands in the seventeenth and eighteenth centuries. Here, novice bards began training outside the school by mastering reading, writing, and grammar—stirring the cauldron of speech until the drops of Awen were ready to leap from it. Then they passed an examination and entered the bardic school for intensive training.

This training took a distinctive form. Students were given a subject and a traditional verse form, and retired to unlit, windowless rooms to work out their poems; one Scottish account describes poets lying with plaids wrapped around their heads and heavy stones resting on their bellies. After a full day or more in darkness and silence, they brought their verses to the teacher for correction or approval. The process was repeated five days a week for the term of the school, which ran, in seventeenth-century Ireland, from autumn to spring equinox. Students attended for six or seven years, then graduated and sought a position as household bard to some noble family.

Taliesin's nine months in the womb of Ceridwen was just such a retreat into silence and darkness, while the gift of poetry ripened within him. The wise sorceress Ceridwen can be seen as the school or the bardic tradition itself—the foster-mother of all true poets. Yet here again there are deeper dimensions, for Taliesin's powers are not simply those of a bard, or even a master bard. He shares an ability to take animal shapes and summon the winds with the Druids of Irish legend, and his radiant brow places him among holy men and women of many other places and times—initiates whose halos, flames, or horns of light symbolize the awakening of the upper psychic centers. Thus Taliesin is a bard, but also more than a bard, and he is initiated not simply into the mysteries of the bardic craft, but into a deeper awakening.

As modern Druids have pointed out more than once, the three stages of Taliesin's initiation take place in any form of spiritual training. There is always a time of patient practice, stirring the cauldron while the potion brews; there is always a sudden awakening into knowledge and power, and the cascade of transformations that follows; and there is always a retreat into silence and stillness while the process of inner development runs its course.

The three phases of enlightenment rarely follow this neat sequence, however. More commonly, spiritual practice tumbles from one to another and back again—one day, slow patient plodding; the next, a sudden leap of insight; the one after that, a seemingly empty time of inward silence, the conscious self flailing about rudderless as deeper levels take care of the work only they can do. Sometimes all three stages happen at once.

The same three stages, finally, define the framework for the system of Druid training taught in this book. As mentioned earlier, this system centers on three core factors, or paths. The first, the Earth Path, challenges you to seek a life more in harmony with nature; here the patient stirring of the cauldron is central. The second, the Sun Path, leads you to celebrate the changing seasons of the year; here the natural world moves through Gwion's shapeshifting flight and sweeps up humanity in its changes. The third, the Moon Path, guides you in the practice of meditation; here the stillness of Ceridwen's womb provides a place in which the hidden depths of awareness slowly reveal themselves.

The legend has many other lessons to teach, and Druid Revival writers have explored some of these over the last three centuries. Still, a legend such as this is an experience to be embraced, not a puzzle to be solved. Let go of the interpretations suggested here, and read back over the tale again. What does it communicate to you? What lessons does it offer?

Arthur, the Once and Future King

Arthur has a much wider reputation than Taliesin—King of the Britons, master of the Table Round, wielder of Excalibur, initiate of Merlin's mysteries, he stands at the pivot of the world's greatest cycle of heroic legends. For hundreds of years, most accounts of Britain's past have had to come to terms with him in one way or another, and three centuries of Druid Revival efforts have woven him and the diverse strands of the Arthurian legend into the fabric of modern Druidry.

Who is Arthur? Most modern historians accept that behind the ornate embroidery of the medieval legend lies the reality of a British military commander in the late fifth and early sixth centuries of the Common Era. The Welsh chronicler Nennius, whose book is the oldest known source to mention Arthur, calls him *dux bellorum* ("war leader") rather than king, and recounts nine battles in which he defended Britain against Saxon invaders. Archeologists and historians confirm that Saxon expansion in Britain came to a sudden halt at the end of the fifth century and took more than fifty years to get started again. That period of successful resistance, unparalleled anywhere else during the fall of the Roman Empire, has been called "the Arthurian fact."

Measured by its historical results, that fact was of immense scale.

Elsewhere in Europe, barbarian invasions led to total collapse as Roman authority went to pieces and nothing took its place. The fifty-year reprieve won by Arthur's battles allowed a shieldwall of Celtic kingdoms—Cornwall, Wales, Rheged, Dalriada, and others—to take shape amid the catastrophic implosion of Roman power. Behind that shieldwall lay Ireland, where monks preserved the heritage of classical civilization and passed it on to the rest of Western Europe in the centuries that followed. Had Arthur failed, the Saxons might easily have swept straight across Britain to the Irish Sea and ravaged Ireland in turn. Without the Arthurian fact, there might well have been nothing to light new lamps in the Western world as the Dark Ages closed in.

The impact of the historical Arthur was more than matched by the growth of his legend. Stories doubtless gathered around the historical Arthur while he still lived, and his death in battle in the year 532 gave the legend unstoppable momentum. In Wales, Cornwall, and Brittany, descendants of the people Arthur defended remained politically independent for centuries and preserved their cultural identity straight through to the present day. Tales of Arthur formed part of that identity, and grew in the telling. Some 600 years after Arthur's time, the tales reached the ears of French and Norman poets, who recognized a gold mine when they heard it and shaped the Arthurian legend into its finished form.

The result was a tapestry woven from wildly diverse sources. Glimpses of the historical Arthur and his struggles with Saxon invaders show through now and again, mostly in early romances such as Geoffrey of Monmouth's *History of the Kings of Britain*. The canons of medieval literary taste play a larger role, with details of court life and knightly combat from the Middle Ages obscuring the context of a Dark Ages warlord. Still, strange things found their way into the romances. Giants, goddesses, wizards, and bizarre quests wedged themselves into the growing legend, and the knights of the Table

Round in their shining armor rode through a landscape dominated by powers of an older world.

William Blake put his finger squarely on this deeper dimension of Arthurian legend in an 1809 essay: "The giant Albion, was patriarch of the Atlantic; he is the Atlas of the Greeks, one of those the Greeks called Titans. The stories of Arthur are the acts of Albion, applied to a prince of the fifth century."[6] What Blake was suggesting is plausible enough: behind the historic Arthur, the "prince of the fifth century," stands an ancient mythology.

In his 1971 study *Camelot and the Vision of Albion*, Arthurian scholar Geoffrey Ashe proposed that Blake's visionary guess was precisely correct. Much of Arthur's legend, Ashe argued, makes no sense as distorted history, but plenty of sense as half-remembered mythology. In the twilight years of British Paganism, myth turned to legend as old gods gave way to new heroes; the process has happened many times before and since. Behind the Arthurian legend, then, lies the Arthurian myth as well as the Arthurian fact.

This shows up all through Arthur's legendary biography, and, above all, in the mystery of his fate. Early sources record his death in the battle of Camlann in 532, but within a short time, strange stories began to circulate. "Not wise the thought, a grave for Arthur," comments an old Welsh proverb. By the twelfth century, people in Brittany, Cornwall, and Wales insisted that Arthur was not dead, but had gone to a hidden isle in the West and would someday return.

This story reaches deep into mythic levels of consciousness and has echoes in the legends of many lands. Classical myths refer to the god Saturn, who ruled a golden age in the distant past before being dethroned by Jupiter, and who slept in an island to the far west; some tales suggested that, like Arthur, he might someday return. The Welsh legend of the titanic Bran the Blessed, whose head lived on for years after his death in an island to the west of Britain, is another working of the same mythic patterns.

119

Insights of this sort have led some scholars to attempt to tease out the original mythology from within the tangles of Arthurian legend, just as other scholars have tried to extract the actual history of the Dark Age chieftain from the same source. From a Druid perspective, these efforts risk missing the value of the legend as it actually exists. Most of these reconstructions simply retell the old stories in the language of today's intellectual fashion, leaving out anything that doesn't fit whatever model the author fancies.

Between history and myth, however, lies the legend itself, in all its complexity and power. The same attitude toward myth discussed earlier is just as relevant to more complex legends. Like the story of the three rays of light, the stories of Arthur are important because of what they teach. Arthur is Everyman, and the knights and ladies of his court are the powers of the soul, its actions and passions. At the same time, Arthur is Arth Fawr, the Great Bear, pivot of the starry heavens, and his Table Round is the wheel of stars; the knights and ladies of his court are vast cosmic powers projected downward onto the screen of human life. The same is true of every great cycle of heroic legend: as above, so below.

As the story of Taliesin provides Druidry with its model of initiation, the stories of Arthur provide it with symbols and narratives for the world into which the initiate enters. The weird ambivalence of Arthurian legend—half of it in the brightly lit world of medieval pageantry, half of it in a wider realm where vast archaic shapes move half-seen in twilight—matches the experience of generations of modern Druids, who live every moment in both the everyday world of industrial society and the greater realm of primal spiritual powers.

As in the legends and in the reality of Druid life, the richness comes from the fusion of the two worlds. Some spiritual teachings attempt to escape from ordinary life into a purely spiritual existence, mirroring the modern secular philosophies that attempt to abandon the spirit in favor of a wholly material existence. Druid tradition supports neither of these unbalanced paths. Rather, sacred and secular, spiritual and material blend constructively in Druid practice, with the additional factor of creativity and the arts turning the binary into a balanced ternary.

The figure of Arthur holds another lesson for Druids as well. The Arthurian legends have their roots in a time when events could have moved in many ways, and one brave leader and his supporters tipped the balance of history. The historical Arthur did not simply live in the world; he helped shape it. It's probably not an accident that modern Druids, whose paths draw deeply on Arthurian lore, have been in the forefront of many progressive causes down through the years. Some have tried to influence the political process in one way or another. More common, and usually more successful, have been efforts by Druids to change the world through the force of personal example, through living the change they hope to make in the world.

The image of kingship, expressed and symbolized in the figure of Arthur, is important to many modern Druids for this reason. This has nothing to do with the tepid reality of the world's surviving monarchies, much less with antidemocratic political stances of the sort that infest some other spiritual traditions. Rather, it represents a deep awareness of the possibilities within each human individual. The English word "king" was originally *cyning*, literally "he of the kin," while "queen," anciently *cwen*, meant simply "woman." The king was *the* member of his kin, the queen *the* woman, not in an exclusive sense, but simply because he or she had the task of embodying the full potential of humanity to his or her people.

Thus, in a Druid sense, every man is a king and every woman a queen, since each human being has the capacity to burst free of the limits of time and place, and express something of the immense possibilities contained within the human body, mind, and spirit. In an age when the collective imagination of humanity stands at a low ebb, when media and the makers of opinion lure the unwary with an image of humanity reduced to a cog in the machinery of consumption, the figure of Arthur reminds Druids that their path embraces the immense challenge of becoming fully and magnificently human.

This is one lesson of the ancient title borne by Arthur and the kings of his line. The word *Pendragon* literally means "head or chief dragon." While the dragon is among the most complex of all mythic symbols, it has two principal meanings in Druid tradition. It represents the hidden energies and potentials within the land, the dragon currents of the Earth, but it also represents the hidden powers and possibilities within the self, the dragon fire that lies sleeping within the core of each individual. An essential part of Druid teaching is that these two meanings aren't really separate, for nature is inside us as well as around us at each moment. Each of us, each living being, is an expression of the life of Earth, part of a greater unity. Our awakening into wisdom brings healing to the Earth, and ultimately brings the Earth itself closer to the fulfillment of its own tremendous destiny.

Merlin, the Mage

Merlin was archmage of Britain and the initiator of the young Arthur. Most people these days who think they know about Merlin see him complete with pointed hat, long white beard, a robe spangled with moons and stars, and a Harry Potter-esque owl on his shoulder, all borrowed from recent media portrayals. A closer look at the original tales reveals a far more complex figure whose legend teaches lessons of great power and wisdom.

Merlin first appears in Welsh sources as the poet Myrddin, author of several enigmatic Welsh poems traditionally dating from the sixth century. Scraps of information preserved in chronicles and poems sketch a dim but extraordinary image: madman, seer, and poet, dwelling beneath a magical apple tree in a forest in what is now southern Scotland, with wolves and wild pigs as his companions. His name resonates far back into Pagan times, perhaps into Druid tradition; according to Welsh sources, the original name of Britain before it was first inhabited by human beings was *Clas Myrddin*, "Merlin's Enclosure."

In the twelfth century, Geoffrey of Monmouth, author of *The History of the Kings of Britain* and *The Life of Merlin*, drew on these traditions (among many other things) to rework the shadowy figure of Myrddin into an archetypal form of immense power. In Geoffrey's tale, the tyrant and usurper Vortigern, one of the supreme villains of British legend, sought to cement his power over the British kingdom by building a stronghold at Dinas Ffaraon in the mountains of northern Wales, but the work of each day fell to the ground each night. Vortigern's magicians told him that the walls would not stand until the blood of a child with no father was sprinkled on the stones. Enter Merlin—Geoffrey's reworking of the Welsh name, probably chosen to avoid a pun on the French obscenity *merde*—born of a Welsh princess seduced by a spirit, who turns the tables on his would-be sacrificers.

Presently, the boy Merlin was brought before the king. He challenged the magicians to explain why the walls fell each night. When they admitted their ignorance, he told the king to have his workmen dig beneath the site, foretelling that they would find an underground pool containing two fighting dragons—one red, one white. The workmen uncovered the pool, and the dragons soared up into the sky and battled each other until the red one conquered the white. Merlin then prophesied Vortigern's death, foretold the return of the rightful king, Aurelius, and spoke of a greater king to come.

All these things duly happened: Aurelius and his brother, Uther, returned from exile; Vortigern was trapped and burned to death in his mountain stronghold; and Uther, with Merlin's help, fathered the future King Arthur on the wife of his vassal, Gorlois of Cornwall. This last event marks Merlin's final appearance in the *History of the Kings of Britain*; *The Life of Merlin* relates his life long after Arthur's death. Here, the image of the mad poet in the forest abruptly replaces the prophet and mage, and images from the ancient Welsh poems take a leading role. Despite its title, the *Life* recounts only a few years late in Merlin's life, as he faces the deaths of friends in battle, flees to the woods in grief and madness, and finds his way back to sanity through a profound process of transformation. Taliesin appears in Geoffrey's *Life* as a friend and younger contemporary of Merlin. In one powerful scene, the two reminisce about their voyage over the sea with the wounded Arthur to an island in the West.

The image of Merlin that became most popular in later years—the wise teacher and counselor of King Arthur—appears nowhere in Geoffrey's account. That came with later writers of Arthurian legend—especially Robert de Boron, whose romance *Merlin* appeared around 1200 and set the tone for nearly all subsequent accounts of the wizard. De Boron's Merlin arranges for Arthur's conception and upbringing, and stage-manages the events that lead the young prince to his destined throne. Child of a demon and a holy virgin, Merlin has powers and habits like those of Taliesin: he knows past and future, changes his shape, and commands the elements; he plays magical pranks on those around him, but displays a fierce loyalty to those he chooses to serve.

Merlin's death is recounted variously in the romances. The most popular version today—his entombment beneath a stone by the trickery of Nimue—seems to have been invented late in the legend's evolution. Older stories tell stranger tales. The Didot *Perceval*, one of the oldest surviving versions of the Grail legend, tells that Merlin could not die until the world's end, and instead retreated into a mysterious *esplumoir* near the hermitage of Perceval, never to be seen again. What an esplumoir might be has been much debated by scholars. The most likely suggestion makes it the cage in which moulting falcons were kept until they regained their feathers—a pun on Merlin's name, which is also the name of a variety of falcon.

Geoffrey of Monmouth's *Life of Merlin* never describes his death, but has him withdraw in old age to an observatory with seventy windows in the depths of the forest, possibly another version of the esplumoir legend. Still another account from Welsh lore claims that Merlin and the thirteen treasures of Britain now wait in a house of glass in the ocean's depths. A dissonant note comes from two accounts in a medieval manuscript now in the British Library that identify a mad wilderness-dwelling figure named Lailoken as Merlin and recount his threefold death by being simultaneously hit with stones, pierced with a stake, and drowned in a stream.

What is one to make of this torrent of myth, folktale, and garbled history? Scholars, occultists, and Druids have wrestled with that question with varying results. Several writers suggest that Myrddin, or Merlin, was a title rather than a personal name, and tease out two possibly historical figures from the legend. The first, Myrddin Emrys, or Merlinus Ambrosius, lived in southern Wales and Cornwall in the late fifth and early sixth centuries C.E. He was teacher and counselor to the historical Arthur. The second, Myrddin Wyllt, or Merlinus Caledonius, lived almost a century later in southern Scotland and retired to a forest hermitage after the death of his patron King, Gwenddolau, in the battle of Arderydd in 573 C.E. The confusion in Geoffrey's books and their sources becomes a good deal clearer with this distinction in place, but it's still merely a hypothesis.

As with the other figures we've examined, Merlin moves at will through a territory in which myth, history, and spiritual teaching blur. That a person named Myrddin played an important part in sixth-century Britain seems likely, although his relation to the Arthurian fact is anyone's guess. That there may be a Pagan god behind much of the Merlin legend also seems likely, although any attempt to recover his symbols and myths strays quickly into speculation. Unquestionably, however, there is a legendary wizard named Merlin, whose complex biography has much to teach modern Druids.

If the legend of Taliesin shows the Druid way of initiation, and the tales of Arthur describe the world in which Druids walk, the stories of Merlin speak of Druids themselves. It's surely no accident that Merlin's distinctive character—acerbic yet benevolent, masterful yet humble, a little frightening yet sometimes a little ridiculous, and always wryly amused by the antics of those around him—appears time and again in the great figures of the Druid tradition. As Geoffrey also reminds us, to the end of his life, Merlin remained open to change and new insight. He never fell into the trap of thinking that he knew it all.

Merlin's relationship with magic raises points that are deeply relevant to modern Druidry. While modern media portray him as a sorcerer, the old legends draw a more complex picture of his powers. In his confrontation with Vortigern, Merlin acts as prophet and seer, not magician; his gift is not power, but rather knowledge of hidden things and meanings. In *The Life of Merlin*, he begins with the same gift and opens up its deeper potentials, slowly developing from seer to philosopher and sage. Even in later literature, from Robert de Boron to Thomas Malory and beyond, Merlin's wisdom and knowledge of the unseen are usually more important than his power to shape the world through incantations and spells.

One lesson this teaches is that magic, in the strict sense of the word, isn't essential to the Druid path. Magic was usefully defined by British occultist Dion Fortune as the art and science of causing changes in consciousness in accordance with will. The elaborate trappings of magic—colors, numbers, geometrical forms, words of power, incense, candles—are simply convenient tools that allow magicians to focus their minds in certain ways and draw on powers of consciousness that modern industrial culture doesn't understand very well. Classical and Irish sources agree that ancient Druids practiced magic, and some Druids today do as well.

Yet the heart of the tradition lies elsewhere. Since the beginning of the Revival, the core element of Druid spiritual practice has been meditation, not magic. Like Merlin himself, modern Druids have made wisdom rather than power the center of their work. This unfolds from the fundamental ideas of Revival Druidry. The Druid quest to understand and move in harmony with nature leads inevitably to a quest to understand our own nature, and meditation holds the key to this quest. The quest for self-knowledge certainly combines well with magical training, but it also combines equally well with other disciplines—the bardic arts of music and poetry, systems of natural healing, and "soft" martial arts such as aikido and t'ai chi ch'uan, all of which are highly popular among modern Druids. Meditation and the quest for self-knowledge remain the living center of the tradition, and open the way to its deepest dimensions.

This lesson is reinforced by the legendary accounts of Merlin's strange fate. Mystical traditions around the world tell of people who remain powers in the inner realms of the human mind and spirit long after withdrawing from physical life. Some of these accounts echo Merlin's tales with curious exactness. Japanese Buddhists of the esoteric Shingon tradition believe that the first great Shingon master, Kobo Daishi, waits in eternal meditation within the sacred mountain, Koyasan, while still mysteriously appearing to those in need who call on him for aid. The legendary founder of the Rosicrucian order, Christian Rosenkreutz, is said to lie deep in contemplation in a hidden vault with seven sides beneath Mount Abiegnus, the Mountain of Fir Trees—a guiding presence for all who follow the Rosicrucian path. Similar accounts surround Hermes Trismegistus, traditionally the first teacher of occult wisdom in the West, and many other figures from myth and sacred history.

Merlin, in his esplumoir of many windows, is another such figure, as is Arthur himself. Legends from across the British Isles claim that Arthur and his knights lie asleep in a cave within some local hill. This cluster of archetypal images suggests again the hidden unity connecting the dragon power of the land with the dragon power of the self—for, of course, Merlin is also associated with dragons, since his vision of dragons rising and battling in the winds above Dinas Ffaraon marked both the awakening of his own prophetic powers and the surging forces within the land that brought the age of Arthur into being.

Yet there's a more straightforward lesson in these same figures. Like all the world's mystical traditions, Druidry teaches that each human being contains potentials beyond anything we can imagine. We are each an esplumoir in which a Merlin waits, a hollow hill in which an Arthur lies sleeping. To recognize this is to take the first great step toward the awakening of vast powers within the silent places of the self.

Further Reading

All the subjects covered in Part II have a sizeable literature of their own, and many students of Druidry end up acquiring a fair collection of books about them. Some books that should not be missed are these:

Geoffrey Ashe, *Camelot and the Vision of Albion* (London: Heinemann, 1971). A seminal work by one of the premier Arthurian scholars of the twentieth century, exploring the mythic dimensions of the legends of Arthur.

Steve Blamires, *Celtic Tree Mysteries* (St. Paul: Llewellyn, 1997). A capable study of the treelore and magical traditions associated with the Ogham fews, with useful practical exercises.

George Calder, *Auraicept na n-Eces: The Scholars' Primer* (Edinburgh: John Grant, 1917). Preserved in the medieval *Book of Ballymote*, this standard textbook of Irish bardic schools preserves most of the surviving Ogham lore. Essential for the Druid scholar.

Robert Lee "Skip" Ellison, *The Druid's Alphabet* (Earth Religions Press, 2003). A convenient handbook of traditional Ogham lore by the Archdruid of Ar nDraiocht Fein (ADF), one of the liveliest of contemporary Druid organizations.

Geoffrey of Monmouth, *The History of the Kings of Britain* (London: Penguin, 1966). The original Arthurian bestseller, this book was first published in 1136. It covers not only Arthur, but an entire mythic history of Britain from its first settling by Brutus the Trojan to the death of Cadwallader, the last Celtic king of Britain. An essential starting place for anyone interested in the Arthurian legends.

Gareth Knight, *The Secret Tradition in Arthurian Legend* (Wellingborough, Northamptonshire: Aquarian, 1983). Based on the work of English magician Dion Fortune and her magical order, the Society of the Inner Light, this book covers the whole sweep of Arthurian legend from the standpoint of Western occult philosophy. Essential.

Sir Thomas Malory, *Le Morte D'Arthur* (New York: Random House, 1994). First published in 1485, this is the classic English retelling of the tales of Arthur. Not to be missed by anyone interested in Arthurian lore.

John Matthews, ed., *The Bardic Source Book* (London: Blandford, 1998); *The Celtic Seers' Source Book* (London: Blandford, 1999); *The Druid Source Book* (London: Blandford, 1996). Three substantial anthologies of sources and essays on ancient and Revival Druidry, with material bearing on most of the topics raised here. Very much a mixed bag, but each volume contains treasures.

Damian McManus, *A Guide to Ogam* (Maynooth: An Sagart, 1991). The academic state of the art in Ogham studies, this volume presents most of what is currently known about the origins, history, and development of the ancient script of the Druids. Some of the linguistic material is heavy going for the ordinary reader.

Paul Rhys Mountfort, *Ogam: The Celtic Oracle of the Trees* (Rochester, VT: Destiny, 2002). The best book currently available on Ogham as a divination system, with a valuable commentary on the twenty primary fews.

Ross Nichols, *The Book of Druidry* (London: Thorsons, 1990). A comprehensive study of Druid lore by one of the major figures of the twentieth-century tradition. A complex and demanding book requiring careful study and meditation; an astonishing amount of lore is packed into it.

O'Boyle, Sean, *Ogam: The Poet's Secret* (Dublin: Gilbert Dalton, 1980). A study of Ogham as a possible notation for music, based on old Irish harp traditions. Rare, but worth finding.

Dell Skeels, *The Romance of Perceval in Prose* (Seattle: University of Washington Press, 1961). A capable translation of the Didot *Perceval*, one of the oldest, simplest, and most evocative versions of the Grail legend.

R. J. Stewart, *The Mystic Life of Merlin* (New York: Arkana, 1986); *The Prophetic Vision of Merlin* (New York: Arkana, 1986); *The Way of Merlin* (London: Aquarian, 1991). A major theoretician and teacher of the modern magical revival, Stewart has made a particular study of the Merlin legends, using authentic medieval sources rather than modern stereotypes. Highly recommended.

Nikolai Tolstoy, *The Quest for Merlin* (Boston: Little, Brown, and Co., 1985). The best introduction to the legendary and historical dimensions of Merlin—readable and highly detailed.

Williams ab Ithel, Rev. J., ed. and trans., *The Barddas of Iolo Morganwg* (York Beach, ME: Weiser, 2004). The core document of the nineteenth-century Druid movement, compiled from Iolo Morganwg's manuscripts. Material on the three rays of light, the three circles of manifestation, and the three Druid elements entered modern Druidry through this vast and disorganized collection.

PART III:

WAYS OF THE SACRED GROVE

Initiation

into the Druid

Tradition

7

THE EARTH
PATH

I N PARTS I AND II, WE TRACED
the origins of modern Druidry and examined the core concepts of its philosophy and symbolism. But learning about
Druidry is one thing; becoming a Druid is another. The first
often leads to the second, but the passage entails a leap
from history and lore to the actual challenges of Druidry as
a living path. You can study the Druid tradition in a
detached, intellectual way, but initiation into Druidry
demands the participation of the whole self. In Part III, we
will explore three paths that lead to the sacred grove of
Druidry: The Earth Path, the Sun Path, and the Moon Path.

Each of the three paths provides a core element of a
Druid way of life. The Sun Path responds to the need for
participation in the great cycles of nature and spirit; the
Moon Path answers the need for self-knowledge and a

way of developing the subtle potentials of the self. The two together offer a balanced system of natural spirituality. Yet spirit must express itself in the world of matter or it accomplishes nothing. Insights from meditation and ceremony gain their full power and meaning when reflected in the details of everyday life. A spirituality based on reverence for nature thus finds its full expression in a life in harmony with the living Earth itself and this is the essence of the Earth Path.

The Earth Path is the path toward living in harmony with nature. Since the beginning of the Revival, Druids have taken issue with the unnatural habits of a civilization alienated from the living Earth. As industrial society moved further out of balance with nature, Druids responded creatively, finding ways to restore the balance in their own lives and communities. Those responses are even more necessary today.

At the same time, the effort to return to a more natural way of living provides fertile soil in which nature spirituality can flower and bear fruit. Mystics and sages have pointed out for centuries that mindless pursuit of wealth, power, and luxury gets in the way of spirituality and, indeed, inhibits ordinary happiness as well. Since "mindless pursuit of wealth, power, and luxury" is a fair capsule description of modern industrial society, and middle-class people in the industrial world today have more wealth, power, and luxury at their disposal than the average medieval king, this bit of traditional wisdom has even more relevance today.

Concerns of this sort have helped create a new movement toward saner lifestyles under the label of "voluntary simplicity." The voluntary simplicity movement, when it hasn't been debased into a sales pitch for yet another round of allegedly simple consumer products, has much to offer the Druid tradition. Its proponents make a compelling case that people in the developed world have cluttered their lives with so many possessions and surrendered so much of their time to buying, maintaining, and paying for them that there's no room left for spirituality, community, relationships, personal development, or simple happiness. They propose replacing the rat race of endless consumption with a life that's outwardly simple and inwardly rich, and have evolved useful tools for sorting out the difference between needs and wants, getting off the treadmill of consumption for its own sake, and moving toward a less cluttered lifestyle. These can be a valuable part of any Druid's Earth Path.

The Druid vision reaches beyond these issues, however, into an intense awareness of humanity's connection with the living Earth. It's important to move away from modern habits of mindless consumption, but it's equally important to challenge the ways of thinking that underlie those habits. The most important of these misunderstandings is the notion that human beings are somehow outside nature, free of its laws and without any need of its gifts. Even in the most straightforward practical sense, this assumption is as misguided as it is arrogant. The air we breathe, the water we drink, the soil and weather that produce our food, and around $33 trillion a year in free services to the global economy—around three times the value of *all* human economic activity—are produced for us moment by moment by natural cycles in the Earth's biosphere.

Yet human connections with nature reach in other directions as well. Until a few centuries ago, nearly everyone lived their lives in intimate contact with nature and its cycles. Our minds and spirits, as well as our bodies, evolved in that context, shaped by the rhythms of the living Earth. Researchers in the new discipline of ecopsychology have shown that human beings are never quite sane when they're isolated from natural environments. Some of the more perceptive writers on voluntary simplicity suggest that our society's frantic search for new sensations, new products, and new lifestyles gets its power from a vague but pervasive sense that we've lost something essential. The

Druid vision suggests that what's been lost is a living connection with nature, as essential to human life as sunlight, rain, and soil are to the life of a tree.

The Earth Path presented here thus has two major aspects: reconnecting with nature through study and nature awareness, and moving toward greater harmony with nature through carefully considered changes in everyday life to remove barriers that separate you from nature and lighten your impact on the living Earth.

These concerns become more serious in the light of today's ecological crisis. Fantasies about infinite material power and wealth are dysfunctional enough by themselves. On a finite, fragile planet, they become forces that push Earth's human and natural systems alike against unyielding planetary limits. Every aspect of the biosphere faces growing strain today, and those strains drive wild swings in planetary systems that impact our lives more disastrously each year. From climate shifts and violent weather caused by global warming, to fisheries and forests ravaged by overharvesting, to new diseases launched into the world by the disruption of tropical environments, natural balances essential to humanity's well-being are spinning out of kilter.

At the same time, many of the resources on which humanity now depends—especially petroleum, the hydrocarbon lifeblood of industrial society—are running dangerously low. There's no shortage of talk about solutions, to be sure, but most of the proposed solutions call for more of the same dysfunctional habits driving the crisis: pump more, burn more, produce more, pollute more. Meanwhile, no one does anything to put the proposed solutions into effect, or even test them to find out if they work on a large enough scale to make a difference. Politicians pursue business as usual and business leaders drag their feet, while problems multiply and options narrow. It's a recipe for a very ugly future.

Many people who are aware of these problems believe they themselves can't do anything about them. Yet the dirty little secret of the ecological crisis is that it results from choices made by ordinary people. To a remarkable degree, the governments and corporations so often blamed for the world's predicament have simply responded to the demands of citizens and consumers. No one forces the average American to use four times as much energy as the average European every year, or the average European to use five times as much energy as the average Chinese. No one forces people throughout the industrial world to run up immense debts and use up irreplaceable natural resources to produce an unending stream of worthless consumer gewgaws. In fact, every political leader in recent years who has tried to interfere with these habits, or even suggest that they ought to be moderated, has been voted out of office in short order.

Thus there's a vast hypocrisy in the "SUV environmentalism" that insists something be done about the ecological crisis, but refuses to consider giving up any of the short-term benefits of ecological destruction. Those short-term benefits include much that people in the industrial world consider part of a normal lifestyle. If human exploitation of the natural world is reined back to a sustainable level, people will have to make do with fewer goods and services, fewer conveniences, and fewer choices. There's no way around that unpalatable fact.

Yet it's not as though we have a choice. By definition, unsustainable lifestyles can't be sustained forever. Like it or not, we face a transition from a consumer society where most work is done by burning irreplaceable fossil fuels at an extravagant rate, to a conserver society where work is done on a far more modest level by sun, wind, water, and human muscle. The only question is whether we make the change voluntarily, while there's still time to save the best achievements of the last 300 years, or wait until ecological troubles and resource exhaustion force desperate last-minute measures amid rising shortages and the chaos of a disintegrating industrial system.

Efforts to force the change to a conserver society through political action, which many people see as the best response to the crisis, have accomplished little, for political change begins with personal change or it goes nowhere. If people refuse to embrace a conserver lifestyle in their own lives, is it likely they'll accept having such a lifestyle imposed on them by political authorities? Nor will vandalism or violence bully people into ecological awareness, as a few misguided people on the fringe of the environmental movement have argued. Such outbursts are the political equivalent of tantrums; beyond alienating people who might otherwise support environmental causes, they accomplish nothing.

The one option left is the hard, but viable, path of personal action and example. Each of us causes part of the burden humanity places on the Earth, and each of us can lighten that burden by our own actions. The changes seem small in global terms, but one person's actions can make a major difference to local ecosystems. The impact adds up as others do the same, and each person who moves toward balance with nature blazes a trail that others can follow.

This isn't simply a theory or a pious hope. The core ingredients of a conserver society—among them organic farming, permaculture, solar and wind energy, and recycling technologies—exist today because a handful of people decades ago decided that the journey back to harmony with nature was worth making, even if they had to make it alone. Some people have dismissed the pioneering ecological work of the sixties and seventies as a failure because it didn't bring Ecotopia overnight, but initiatives launched in those years played out spectacularly in the following decades. Organic farming, the interest of a handful of activists in 1970, has grown into an industry that earns billions of dollars and keeps millions of acres free of toxic chemicals each year. Recycling has followed a similar growth curve, from the cause of a few ecological radicals to a normal part of daily life for tens of millions of people.

Such changes have immense potential, for the transition to a conserver society also has to begin with the personal dimension. Each person who moves toward a conserver lifestyle takes the world one step closer to a viable future. If industrial society comes to its senses and reshapes itself to work within the Earth's limits, those who already live in harmony with the Earth can play a central role in helping others respond creatively to the task. On the other hand, if industrial society holds its present course until it crashes into the consequences of its own arrogant mistakes, those who have made themselves less dependent on it will be more likely to survive the impact and better prepared to rebuild a sustainable world afterward. If things fall somewhere between the extremes, as seems most likely, those furthest along the path to a sustainable future will again be best prepared to face the perils and possibilities of the transition.

Here again, however, the situation demands awareness as well as action. Too many would-be environmentalists know little or nothing about the natural world they think they're defending. As the environmental crisis grows more severe and options become fewer, an understanding of nature combining intuitive insight, factual knowledge, and personal experience becomes one of the few worthwhile sources of guidance. Thus the Druid tradition, like other nature-centered spiritual paths from traditional cultures around the world, has special gifts to offer during the difficult changes we face in the present and the future.

Nature Awareness

A crucial part of the Earth Path involves waking up to the presence of living nature in your own life. Many people in the modern industrial world go through life with their bodies surrounded by a cocoon of technology and their minds flooded with perpetual chatter from the media. Living and working in climate-controlled buildings, with artificial lighting to see by, commercial music to hear, synthetic scents to smell, chemically flavored foods to taste, and a completely manufactured environment to touch, it's no wonder so many modern people are deluded into thinking of nature as an unnecessary luxury, and fail to notice that their glittering artificial world depends, moment by moment, on vast inputs of materials and energy wrenched from their places in the cycles of the living Earth.

Psychologists have shown that many kinds of mental illness have a central factor in common. Despite the stereotype, mentally ill people don't lose the ability to think clearly; psychotic delusions are often masterpieces of internally consistent deductive logic. The trouble comes because this elegant reasoning loses touch with anything outside itself. Deductions become delusions when they stop being tested against the way the world actually works.

In the modern world, we face the same problem on a global scale. The artificial environment in which so many people spend all their time is entirely shaped by human minds, obeying rules unrelated to the way things work in the natural world. This distorts our thinking about countless issues. Since plumbing and garbage services take our waste out of sight, we lose track of the fact that waste never just "goes away"—it always ends up somewhere. Since money controls the flow of goods and services within our manufactured world, we lose track of the fact that all the money in the world won't let us violate the laws of nature. It's not going too far to say that we live inside a delusion turned real, a hallucinatory fantasy world frozen into material form. Yet the way out of the hallucination is as close as nature.

The word *nature* literally means "that which is born." When the poet e.e. cummings spoke of the difference between "a world of made" and "a world of born" in one of his most famous poems, he gave voice to many of the themes of this chapter. Druidry begins with trust in nature— recognition that the natural world is our home, our source, and the teacher of the wisdom we most need to learn. From a Druid standpoint, everything that comes from nature, everything that's born rather than made, provides us with a desperately needed reality check.

That reality check is embedded everywhere in the world around us. There's a real value to spending time in unspoiled wilderness places far from people and their creations, surrounded by living nature on all sides. Human beings evolved in such settings, and we regain a link with the Earth and a sense of perspective from renewing our contacts with nature in this way. Still, the idea that nature exists only in isolation from humanity is misleading at best, and can turn into yet another excuse for mindless consumption. Treating nature as raw material for human recreation or personal retreat is still treating nature as raw material.

Many places around the world are co-creations of humanity and nature, places in which the made and the born dance together and create beauty and healing. Rural areas still unravaged by industrial agriculture are full of such places, where fields, pastures, hedgerows, coppices, and orchards quilt the landscape, and houses of natural materials nestle close to the land. Wild animals and plants make their homes in these landscapes, and most traditional cultures guide the interface between wild and tame in ways that benefit both. It's possible to get as close to nature in such a place, and learn as much, as in a wilderness.

Even in the midst of modern cities, nature is present. Outdoor educator Tom Brown Jr., in his excellent *Field Guide to the Forgotten Wilderness*, describes encounters with scores of wild creatures in downtown New York City and other urban centers. During two decades living close to Seattle's urban core, I spent time watching urban raccoons, opossums, deer mice, wood rats, and turtles, and listened to the calls of unseen coyotes in city parks. Birds, from tiny wrens and fox sparrows up to great blue herons and bald eagles, also brought nature closer to me in the city. A patch of garden not much bigger than an average living room graced my life with dozens of wild herbs ("weeds") alongside the plants I chose for it, and brought five different kinds of bees, each with its own habits, to nourish their hives and my spirit. All these creatures have lessons to teach the Druid.

The door to nature is thus much closer than it looks, and the hallu-cinatory unreality of modern industrial culture is at least as much a matter of selective inattention as anything else—the "mind-forg'd manacles" William Blake described in his poetry. It takes time, patience, and effort to break through that habit of inattention, but the results repay the costs many times over.

Stillness, Focus, and Study

Three practices make up the nature awareness phase of the Earth Path: stillness, focus, and study. Stillness requires that you slow down to nature's pace. At least once a week, go to a place full of nonhuman liv-ing things, find a place to sit, and spend at least ten minutes sitting still and noticing everything around you. Relax your body, and turn your attention away from the chatter of thoughts and worries that usually fills your mind. Sit still, watch, listen, and feel.

The first few times you try this, you may find yourself bored or frus-trated. You may expect something dramatic to happen, and become dis-appointed when it doesn't. This happens because your mind is still running at the frenetic rate of modern industrial society and not at nature's slower pace. Simply work at paying attention to the natural setting around you. Try to broaden your awareness—notice sounds, scents, wind, light and shadow, changes in the sky and in the distance around you. Let the world fill your mind and guide it.

Go to the same place sometimes, and to different places at other times. If you travel to other regions, find places there where you can become still and pay attention to the world of nature unfolding around you. Gradually, you'll find that the practice of stillness teaches you more than you could learn about Druidry from any other source.

A powerful addition to this practice is the technique of "splatter vision." Splatter vision is the technique of letting your field of vision spread out as wide as possible, instead of focusing tightly on something in front of you. Let your eyes soften and take in everything within your field of vision—to left and right, but also above and below. Scientists have found that the parts of the eye that pick up images toward the sides of the field of vision have proportions of rods and cones—the two types of cells that receive visual images—different from the part that picks up images from straight ahead. Using these rods and cones thus allows you, literally, to see the world in different ways. "Splatter vision" takes advantage of this. Experiment with it in your stillness practice and see where it takes you. The practice of stillness gradually merges into receptive meditation, one of the types of meditation taught below in the Moon Path (see p. 229).

Focus, the next phase in nature awareness, involves paying close attention to what you encounter. You can learn this by focusing your mind and senses on the details of specific natural things: a stone, the bark of a tree, a handful of dirt, a pool of water, the crumbling end of a fallen log. As with the practice of stillness, this requires time. Choose something and spend at least five minutes as close to it as possible, with every sense focused on it. Let it be the center of your world for that period of time. Push aside all other concerns; simply look, listen, smell, and feel.

Focus practice quickly reveals hidden realms of being. Spend ten minutes on your belly on the lawn, paying attention to the little lives that inhabit the microcosmic jungle of grass and lawn weeds, and you'll never look at a lawn in quite the same way again. A magnifying lens is worth bringing on lawn safaris and many other focus expeditions, since sheer minuteness hides entire worlds from ordinary human sight.

Plan on practicing focus on some natural thing at least once a week during the year of your Druid initiation. For the first few months, choose a different target for focus each time. After that, if you wish, you can revisit one or more of your earlier targets and see whether it's changed—or whether you have.

Druidry is a path of knowledge as well as a way of awareness, however. While ordinary learning won't take the place of personal encounters with the living Earth, it has its own importance. Hundreds of years of careful scholarship and study of the natural world have gone into creating today's nature sciences, and these provide important resources for the Earth Path. Thus study is the third step in expanding your awareness of nature.

Almost any source of information about nature can be useful. Many Druids have found, however, that books and resources on the natural history of their own area are especially helpful, because they bring study together with experience. Paying attention to nature takes on new dimensions when you recognize the birds around you, the cloud types above you, or the varieties of stone and soil beneath your feet. When you plan on traveling to another area, take the time to learn something about the living things, geology, and weather there.

It's also worthwhile to follow your own personal interest, reading up on whatever aspect of nature seems most intriguing to you. On the other hand, a certain degree of completeness has its uses as well. Concepts from Druid philosophy can provide a workable framework for your studies. *Annwn*, for example, the primal cauldron of elemental matter, divides into the three Druid elements calas, gwyar, and nwyfre. Calas in the elemental realm takes the forms of stone and soil;

gwyar takes the form of water; nwyfre takes the form of air. Books and resources on geology, geography, oceanography, hydrology, weather, and similar subjects focused on your local area provide the best sources for study here.

Likewise, *Abred*, the circle of incarnate life, can be divided into the three Druid elements. Calas in this circle is plant life, gwyar is animal life, and nwyfre is human life. Books on natural history, botany, animal life, and anthropology focused on your local area, as well as manuals of plant, bird, fish, and animal identification, form your basic resource list in this circle.

Gwynfydd, the circle of spiritual life, embraces the philosophical and spiritual dimension of nature studies. Calas in this circle can be thought of as nature science—ecology, systems theory, and the environmental sciences. Gwyar is nature mythology, especially the traditional myths and legends of the area in which you live. Nwyfre is the field of nature spirituality, both traditional nature-centered religions and recent insights born out of today's environmental crisis.

How much study you manage during the year of your Druid initiation will depend on the time and resources you have available. One book on each of the three elements in each circle, or nine books in all, may be a reasonable target for the average student.

Living in Harmony with Nature

Knowledge and experience form the foundation of the Earth Path, but constructive action defines much of its course. Since profligate lifestyles in the developed world are the primary force behind the destruction of natural ecosystems, there are plenty of opportunities for action and few excuses for unthinking business-as-usual.

Different Druids lead different lives, and no one way of life can or should fit everyone. As William Blake wrote: "One law for the lion and the ox is oppression."[7] Still, everyone can lighten the burden they place on nature. Every part of daily life offers chances to go easier on the Earth, and the consequences of each change add up.

Druids past and present have expressed their commitment to a life in harmony with nature in many ways. Some abandon a modern lifestyle entirely and live in rural areas, growing much of their own food, drawing water and gathering wood, and supporting themselves through some craft or profession in harmony with the living Earth. Others live in more ordinary settings, but find ways to make their way of life as Earth-friendly as possible. Still others find a position somewhere in between, living much of the time in ordinary society in an ecologically sound way, but creating a place in some rural area as a sanctuary for wild nature and a space for personal retreats and spiritual work. All of these are valid approaches.

The widely respected Canadian ecologist Dr. David Suzuki has worked out a list of ten simple ways people in the developed world can sharply reduce their impact on the Earth, and he challenges people to adopt at least three of the ten steps. Over eleven-thousand people have taken up the Nature Challenge so far, including a growing number of Druids. Below is his list of suggested actions:

1. Reduce home energy use by 10 percent
2. Choose an energy-efficient home and appliances
3. Replace dangerous pesticides with alternatives
4. Eat meat-free meals one day a week
5. Buy locally grown and produced food
6. Choose a fuel-efficient vehicle
7. Walk, bike, carpool, or take transit instead of driving
8. Choose a home close to work or school

9. Support car-free alternatives

10. Learn more and share with others[8]

Suzuki's ten steps are well within most people's reach. Most Americans can cut more than 10 percent of their home energy use, for instance, simply by turning off lights when no one's using them, putting insulation on the hot water heater, and wearing a sweater instead of turning up the heat on cool nights—not exactly an arduous lifestyle change. Several other steps require nothing more than keeping the Earth in mind when making a major purchase; the average SUV burns nearly twice the gas and produces nearly twice the air pollution that a station wagon with the same number of seats does, so even soccer moms can have a positive impact.

A longer list of similar suggestions and challenges would be easy enough to create. In fact, later in this chapter I'll do just that. Still, lists of this sort have their limits. Depending on the circumstances of your life, some items may not be practical, while other things not on the list may do at least as much good for the Earth.

This phase of the Earth Path isn't primarily about following a set of rules, in other words. Instead, it's about becoming aware of the ways your actions affect the natural world and taking action accordingly. Every day, you make choices that affect the environment. Keep that in mind as you make your choices and, where possible, choose the option that does the least damage to the Earth.

To do this, you need to learn to think like an ecologist—or a Druid. The science of ecology and the Druid tradition both look at the overall picture, seeing things in context and noticing connections that narrower viewpoints often miss. The following questions can help you see these contexts and connections in your own life:

What is it made of? Everything you use has a source. Products don't simply appear on store shelves by magic, and raw materials don't appear at the factory gates by magic, either. Every material thing you bring into your life is made of things that were once part of the natural world. How they were taken out of nature makes a difference. *Example*: Ordinary paper products are made by cutting down living trees and grinding them into pulp; recycled paper products are made from paper waste that would otherwise go into landfills. When you buy recycled products, you save trees and landfill space at the same time.

How is it made? Everything you use was made, grown, mined, harvested, or gathered by someone, using something. Some methods, tools, and substances used to produce things damage the environment more than others, and this makes a difference. *Example*: Ordinary industrial farms pour hundreds of gallons of pesticides, weed killers, fertilizers, and other artificial chemicals onto each acre of farmland. These end up contaminating soil, groundwater, and the food you eat. Organic farms use compost, companion planting, beneficial insects, and other Earth-friendly techniques to grow crops instead. When you buy organic products, you help keep dangerous chemicals out of the ecosystem.

How far did it travel? Everything you use, unless you make it from raw materials at home, came from somewhere else. Products don't move themselves, and moving them from place to place burns fuel and creates pollution. How far a product was shipped from the place it was made or grown to the place it's used makes a difference. *Example*: Food sold at local farmers markets comes from nearby farms. Food sold at grocery chain stores often comes from hundreds or thousands of miles away. The average item on an American dinner plate has traveled about 1500 miles—the distance from New York City to Houston, Texas, or from London, England, to Athens, Greece—

between field and table. When you eat food grown closer to home, you use less fossil fuel and produce less air pollution.

What inputs does it require? Everything you use needs inputs to make it useful or keep it working. Food needs cooking, clothes need cleaning, cars need gas, appliances need electricity. What you need to add to something, and how much you need to add, makes a difference. *Example*: Gas-powered lawnmowers burn gasoline and produce air pollution. Push mowers, powered by the person using them, burn calories and produce exercise. When you use things powered by sun, wind, water, or muscle instead of things powered by fossil fuels, the Earth benefits—and much of the time, so does your health.

How much waste is involved? Everything you use involves some waste, but how much depends on many factors. Many bad habits of modern marketing and everyday life add to the total waste people in the industrial world produce, and this makes a difference. *Example*: Most modern products come in far more packaging than they need, and making, processing, shipping, and disposing of the packaging adds substantially to the total burden placed on the Earth. It's often possible to buy products in bulk or in minimal packaging. When you buy the product without the packaging, you lose nothing and nature reaps the benefits.

How long does it last? Everything you use has a lifespan from the time you get it to the time you have to replace it. Every time you have to replace something, the natural world has to deal with the impact of disposing of the old item and making, processing, and shipping the replacement to you. All of this makes a difference. *Example*: Paper towels, paper napkins, and disposable diapers have to be thrown away and replaced after one use. Towels, napkins, and diapers made of cloth can be used hundreds or thousands of times. When you use durable products instead of disposable ones, you take a burden off the Earth.

Where does it go? Everything you use ends up somewhere; there's no such place as "away." Every material thing you discard returns to the natural world in some form. How it finds its way back to nature, and what it does when it gets there, makes a difference. *Example*: Food scraps that go into the garbage end up in landfills where they feed rats and help spread disease, or in waste incinerators where they're turned into air pollution. Food scraps that go into a composter improve soil fertility and help plants grow. When you compost food scraps instead of throwing them away, you turn a potential pollutant into a resource.

How necessary is it? Not everything offered for sale actually fills a real need or has a useful purpose. Far too much energy and resources are wasted producing trinkets that get all their apparent value from clever marketing. If people stop buying such things, manufacturers will stop making them, and this makes a difference. *Example*: Many of the products that clutter store shelves these days could sit there gathering dust forever without anyone's life being less complete. When you only purchase things that fill an actual need or add something worthwhile to your life, you stop a waste of natural resources and you free up your own time and money for things that matter more to you.

By asking questions like these and working out the answers, you begin to see your own life in its ecological context. The insights that come out of this process of questioning can guide you toward changes that lighten the burden your lifestyle places on the Earth. Many other Druids have walked this path before you. The twenty steps below describe some of the things they've done. If you've studied the Ogham material in chapter 5, the arrangement of items in the list may make additional sense to you.

Twenty Steps to a More Natural Life

1. If you have room for a garden, or can join a P-Patch or allotment program to garden on public land, grow some of your own food using organic methods.

2. Buy organic, recycled, and other Earth-friendly products instead of conventional ones, even when they cost more.

3. Set the heat ten degrees cooler and the air conditioning ten degrees warmer, replace high-wattage light bulbs with efficient ones, and make a habit of turning off anything that doesn't actually need to be on.

4. Improve your home's heat efficiency by adding insulation, installing insulated window coverings, weatherstripping doors, and putting gaskets behind electrical outlets.

5. Put flow restrictors on your faucets and showerhead to save water. If possible, replace existing toilets with low-water or composting models; otherwise, a half-gallon jug full of water in the toilet tank will limit the amount used in each flush.

6. Never buy anything on impulse. If you think you want something, wait at least twentyfour hours and see if you still want it then.

7. Plant trees whenever and wherever you can, and tend and water them until they can survive on their own (see below).

8. Take a hard look at the electric or gas-powered devices you own. How many could you replace with low-tech equivalents, or simply get rid of? Gather up any that can be replaced or discarded and donate them to charity.

9. Contact your local water, electricity, and heating fuel utilities to find out what conservation programs, rebates, and incentives they offer, and make use of them.

10. Shop at a local farmers market or join a community-supported agriculture (CSA) program, in which you buy a season's worth of produce in advance from a local farmer, and get a box of fresh vegetables and fruits once every week or two all through the growing season.

11. Learn how to entertain yourself and your family and friends instead of letting an energy-wasting machine do it for you. Television and computer games are no substitute for life!

12. Instead of a grass lawn, landscape with plant species native to your area. Local conservation groups can tell you which plants support native butterflies and birds.

13. Whenever you possibly can, walk, bicycle, carpool, or take public transit instead of driving a car. Many people who do this find they don't actually need to own a car at all.

14. Replace chemical cleansers, laundry detergents, and garden compounds with natural or biodegradable equivalents.

15. Take care of as much of your everyday health-care needs as you can using natural methods. Modern medicine is among the most wasteful and polluting of all industries.

16. Live as close as possible to work or school so that you minimize the time and energy wasted in commuting.

17. If you're building a home, include as many Earth-friendly elements in its design and construction as you can. Earth-sheltered construction, passive solar space heating, graywater systems, composting toilets, and other green technologies can make a home into an ecological oasis.

18. Compost all your yard waste and vegetable kitchen scraps in a composter or worm bin, and return the compost to the soil.

19. Recycle everything you possibly can. If your area has curbside recycling, use it. If not, find out what recycling services are available and use them.

20. Donate old clothes, housewares, and appliances to charity, or find other uses for them instead of throwing them away.

Many people have reshaped their entire lives using simple methods like these. The task you face in your year of Druid initiation is a good deal less demanding. For the second phase of your Earth Path work, simply make three changes in your life to move it closer to nature and lighten the burden you place on the Earth. Then follow through on them consistently for a year.

Three changes and no more may seem overly modest, but there's plenty of hard experience behind it. Many people who try to reshape their entire lives in a single rush fail at it, and fall back into old patterns. Those who take one step at a time go more slowly, but more surely. After your initiatory year is finished, you can review the options and your own experiences, and take further steps toward a life in balance with the Earth.

The relatively modest nature of this phase of the Earth Path is meant, among other things, to counter the form of self-righteousness that treats success in some worthwhile task as an excuse to browbeat other people into doing the same. You'll notice that all the options listed above involve changing your own behavior, and perhaps that of your family or housemates if they agree. None of them require you to get your neighbors or the rest of the world to change theirs. This is deliberate, and an important lesson in Druidry.

Becoming a Druid involves taking on the responsibility to bring your own life into harmony with nature. It doesn't give you the right or the duty to tell other people what to do. As Druids, our tradition calls on us to lead by example, not to point the way with empty words. Just as Druid rituals enact the cycle of the year as a pattern of balanced change that human life can follow, Druid ecological action enacts a healthier way of living with the Earth to show other people that it can be done, and how it can be done. As Druid spirituality begins in personal encounters with the symbols and experiences of the tradition, Druid ecological action begins in personal encounters with the challenges of leading a natural life in a radically unnatural society.

Be aware that this approach will land you in conflict on occasion. There are people for whom all environmental action takes the form of trying to make other people do something. In many cases, this is an honest response to difficult situations in which the need for change is desperate and the power of personal action and example seems too slow. On the other hand, insisting that other people do something about the environment is often a way of ducking personal responsibility or avoiding the need to make changes in your own life.

Furthermore, supporters of many unrelated causes have become very skilled at using the language of ecology as bait for the unwary, insisting that, whatever their crusade, it's good for the environment. Even publicists for the nuclear industry have taken to claiming that nuclear power plants are good for the environment, since nuclear power doesn't cause global warming. Since nuclear power produces radioactive wastes so lethal they have to be kept isolated from the biosphere for a quarter of a million years, this is rather like saying that shooting yourself in the head is good for your health because it keeps you from dying of heart disease.

Claims this dishonest are rare, but even idealistic movements have fallen into the habit of making ecological sales pitches based on misinformation. Many people in the vegetarian movement, for example, insist that raising animals for meat is bad for the environment by definition, and anyone who cares for the Earth should adopt a completely vegetarian diet. The idealism and concern for the planet that motivate many vegetarians deserve respect, but their grasp of ecology is less impressive. This sort of "one size fits all" nonsolution can actually cause more problems than it solves. Vegetarians are quite correct to point out that industrial animal raising causes ecological damage, but they apparently haven't realized that industrial vegetable and grain production have an even greater impact on ecosystems, destroying soil and dumping large volumes of toxic chemicals into the biosphere.

In fact, mixed organic agriculture that combines animals and plants into a balanced system benefits local ecologies more than a plant-only approach, since animal manure plays an important role in cycling nutrients through to plants and soil organisms. In arid and semiarid country, furthermore, food plants can't be raised at all without irrigation, which concentrates salt in the soil, and clearing native vegetation, which causes massive soil-erosion problems. In such areas, eco-logically sensitive animal raising that takes the carrying capacity of the land into account is less damaging to the environment than vegetable or grain farming.

Does this mean that if you personally prefer a vegetarian diet and find that your body thrives on it, you should give it up? Of course not. Like all living things, people have different needs, and it doesn't help the Earth or anything else to force everyone to live in exactly the same way. What it means is that a vegetarian diet isn't a panacea for the environmental crisis. In the same way, you're likely to meet people who insist that supporting some political cause, backing some social reform, or following some religion or spiritual path is an infallible way to save the Earth, when the actual ecological benefits of that action may be modest at best. Whenever people insist that their pet crusade is a universal solution to the ecological problems of modern society, you can be fairly sure the crusade means more to them than the environment does.

This is true of Druidry as well. The mere fact that you've decided to become a Druid doesn't do the living Earth any good at all. Druidry only becomes relevant to today's ecological crises when it leads to changes in your life that lighten the burden you place on nature and to actions that help heal the ecosystems around you. Without that, it's nothing but words.

An essential lesson in all the world's wisdom traditions is that good intentions are not enough. If you seek to walk the Earth Path as a Druid, you must take constructive action based on personal knowledge. This is why you're asked to study books on ecology and natural history, and why you're asked to spend time outside, learning the Earth's ways firsthand. Finally, it's another reason why you're asked to concentrate on making changes in your own life, so that you experience their challenges and difficulties as well as their rewards.

Planting Trees

Since early in the Revival, Druids have not only been revering trees, but planting them as well. It's no accident that one of the first English books on tree planting, a 1734 volume on growing oaks by the enthusiastic botanist and gardener James Wheeler, was titled *The Modern Druid*.

It's hard to imagine any action that does more good for the Earth, local ecosystems, and the human community than planting a tree. A mature tree produces as much oxygen as ten people need to breathe, while soaking up carbon dioxide (the most important greenhouse gas) and scrubbing many pollutants out of the air. The same tree can pump hundreds of gallons of water out of the ground on a hot day, releasing it from its leaves as water vapor, helping to moderate temperatures and ease drought. Tree roots bind the soil together, fighting soil erosion. Trees anchor local ecosystems, providing food, shelter, and other resources to animals and birds and improving conditions for other plants by enriching the soil with their fallen leaves.

Planting a single tree may not seem like much, but each tree planted and tended through its early years makes a sizeable contribution to the health of the Earth. If you can possibly do so, plant at least one tree during the year of your Druid initiation. Here's how it's done:

Get a tree suited to your location. Local organic nurseries are a great help here. In the United States, the National Arbor Day Foundation (100 Arbor Ave., Nebraska City, NE 68410, www.arbor-day.org) will happily assist. Other countries have their own tree-planting programs, which should be consulted if you live outside the United States. Climate, sunlight, and the amount and quality of the soil are important factors. So is height; if there are power or telephone lines above, plant a smaller tree. Check drainage by making a hole ten inches deep with a shovel and filling it with water. If it takes more than six hours for the water to drain away, most trees planted in that spot will drown.

Make necessary arrangements. Anywhere that winter temperatures go below freezing, plan on planting in the spring after the ground thaws, but before trees come into bud, or in the autumn after leaves have fallen. If you're far enough south to avoid frost, on the other hand, you can probably plant any time. The day of the New Moon is considered a good day for tree planting in many Druid traditions. In cities and suburbs, contact the utility company several days before you start digging the planting hole to be sure you don't hit a sewer pipe or an electrical conduit.

Prepare the tree. Most trees are available as saplings a few feet tall, with their roots either bare or contained in a "rootball" of dirt. If the roots are bare, soak them in a pail of water for two or three hours before planting. If you have a rootball, water it thoroughly.

Dig the hole. Make it wide and shallow enough so that the "collar" or "root flare"—the place where the trunk turns into the rootstock, clearly visible on nearly all trees—is a couple of inches above ground level. Make it wide enough so that all the roots can be stretched out to the sides without hitting the edge of the hole. This is usually three to five times as wide as the rootball. Pile the soil from the hole to one side. If the soil looks barren, mix a little well-rotted compost with it. Otherwise, leave it alone; fertilizers, even natural ones, should never be used when a tree is planted, since they can burn the young tree's roots.

Put the tree in the hole. Make sure it stands straight—check it from several directions. If your tree came with bare roots, stretch the roots gently out to the sides. If it has a rootball, don't disrupt it; the larger roots have millions of tiny root hairs that can break off with the dirt if you do. Once the tree is settled, fill soil around it, covering all the roots, and pack the soil down gently. Put in more soil than you think you need, since it will settle once you water it.

Water the tree at once. Soak it thoroughly so that the soil settles in around the roots and the tree has enough water to carry it through the shock of being transplanted. Don't compact the wet soil by stepping on it, however.

Perform the tree-planting ritual given below to bless the tree and help it connect to the powers of the cosmos. Like human beings, trees grow along a vertical axis that connects overworld and underworld. They mediate the energies of Aud and Aub to create Aur, the lunar energy of growth and fertility. Just as watering the tree helps the roots make contact with the soil, blessing the tree helps it make contact with the subtler realms of being.

Mulch the area beneath the tree. In natural woods, a thick layer of fallen leaves and organic litter blankets the soil and plays a crucial part in soil ecology. You can get the same effect by spreading a mulch of composted leaves or shredded bark under the tree. Don't let the mulch lie up against the trunk itself, since this can foster diseases; keep it back two or three inches (5–8 cm) from the trunk.

Water at regular intervals. A newly planted tree needs five to fifteen gallons (20–60 liters) of water a week. Unless heavy rainfall does the job for you, you need to provide this for the first two years whenever the ground isn't frozen. Water about three times a week, soaking the ground thoroughly each time. It's often useful to build up a raised ring of soil around the tree, two or three feet out from the trunk so that water poured onto the ground near the tree pools and soaks in rather than flows away.

Keep tending the tree as needed. If the soil is poor, rake up the mulch in the fall, put down a layer of good organic compost two or three inches (5–8 cm) thick, and replace the mulch. Rainwater will carry the nutrients down to the roots. If the tree is damaged, trimming off broken branches will minimize the risk of disease. Your local organic nursery can give you pointers on the area's tree diseases and pests and point you toward solutions.

A newly planted tree will usually need five years of tending to ensure its survival. It will need water through the driest part of each year, treatment for injuries and plant diseases, and new compost at intervals if the soil isn't good. After five years, if all goes well, the roots will go deep enough to find groundwater and reach wide enough to draw in nutrients from a large area. Once this happens, your tree stands a good chance of living out its lifespan without further assistance.

The following tree-planting ritual is traditional in the Ancient Order of Druids in America; you may use it or create one of your own:

> Plant the tree and water it, leaving a little extra water in the pitcher or bowl.
>
> Trace the invoking symbol of the three rays of light \ | / over the upper part of the tree so that the central ray follows the line of the trunk and say: "May the powers of the natural world come together at this point."
>
> Pour the water down the trunk of the tree and say: "May nwyfre, the fire of life, and gwyar, the water of life, descend into the physical form calas."
>
> Trace the evoking symbol of the three rays / | \ over the lower part of the tree, with the central ray again following the line of the trunk. Say: "Within this tree is the fire of growth, Coel. May Ana, the dark Earth, feed it and water nourish it so that Niwalen may break forth in leaf and blossom in due time."
>
> Chant the word *Awen*, "Ah-oh-en," drawing each syllable out and letting your voice resonate. This completes the ceremony.

Further Reading

Among the most important books you can read for your Earth Path work are field guides and handbooks covering your local geology, weather, flora, and fauna, as well as the traditions of local native peoples. You'll need to find these on your own, though local libraries, natural history museums, and nature societies can usually help. Books that will help you walk the Earth Path wherever you live include the following:

Tom Brown, Jr., *Tom Brown's Field Guide to City and Suburban Survival* (NY: Berkley, 1984); *Tom Brown's Field Guide to Nature Observation and Tracking* (NY: Berkley, 1983); *Tom Brown's Field Guide to the Forgotten Wilderness* (NY: Berkley, 1987). Anyone who can go into the Rocky Mountain wilderness with nothing but a knife and come out a year later, healthy and well fed, without harming the land on which he lived, has something worthwhile to say about nature. Tom Brown Jr. did exactly that, and went on to become one of America's most respected teachers of outdoor skills. These books present not only techniques but a way of harmony with the Earth that resonates deeply with Druidry.

Charles Cook, *Awakening to Nature* (New York: Contemporary, 2001). A thoughtful and inspiring guide to reestablishing contact with the world of nature by a widely respected outdoors educator, full of practical exercises as well as wider insights.

Duane Elgin, *Voluntary Simplicity* (rev. ed. New York: William Morrow, 1993). The book that launched today's voluntary simplicity movement, and still among the best guides to "a way of life that is outwardly simple, inwardly rich."

Eugene Kinkead, *Wildness Is All Around Us: Notes of an Urban Naturalist* (New York: Dutton, 1978). Essays and experiences on the presence of wild nature in the city, well worth reading for urban Druids.

Aldo Leopold, *A Sand County Almanac* (New York: Oxford University Press, 1949) One of the enduring classics of environmental thought, essential for the modern Druid.

Georgene Lockwood, *The Complete Idiot's Guide to Simple Living* (Indianapolis; Alpha Books, 2000). Simple, straightforward, and easy to use, this is probably the best single introduction to voluntary simplicity on the market.

James Lovelock, *Gaia: The Practical Science of Planetary Medicine* (Oxford: Oxford University Press, 2000). The single best nontechnical introduction to the ecological sciences in print, by a pioneer of planetary ecological studies. Extremely valuable from the Druid perspective.

Jim Nollman, *Spiritual Ecology* (New York: Bantam, 1990). A guide to the inner dimensions of today's ecological crisis, and attitudes and approaches that may help get us past it. Valuable.

Arthur Plotnik, *The Urban Tree Book* (New York: Three Rivers, 2000). A good general guide to trees in the city, with a very thorough section on books, Web sites, and other resources for the urban tree lover.

Theodore Roszak, *Where the Wasteland Ends* (Garden City, NY: Doubleday, 1972). A powerful and visionary book on the relationship between the natural and spiritual realities ignored or suppressed by modern industrial society. Essential reading.

David Shi, *The Simple Life: Plain Living and High Thinking in American Culture* (New York: Oxford University Press, 1985). Voluntary simplicity has been a theme in American culture since long before the invention of today's frenetic consumer economy. Shi's survey is a good resource for connecting with this forgotten history.

Henry David Thoreau, *Walden and Other Writings* (New York: Modern Library, 1992). The classic on simple living close to nature, Thoreau's Walden has guided more people back to a saner relationship with the Earth than any other book. Not to be missed.

8

THE
SUN PATH

THE SECOND PART OF THE TRIAD
that forms the core of Druid initiation and practice is the
Sun Path, the way of ritual and celebration of the cycle of
the year. The symbolism of the Sun Path reflects an
important aspect of Druid spirituality. Since early in the
Revival, tradition has held that Druid ceremonies should
take place whenever possible "in the face of the Sun, the
eye of light"—that is, out in the open air and the light of
day. Weather and other factors don't always permit this,
of course, but the habit is worth preserving whenever circumstances allow. Druid ceremonies draw on the deep
reverence for nature central to the Druid tradition, and
nature is best encountered in person, even when this
means braving a little wind or rain.

The Sun is a central presence in Druid ritual for another reason, of course. The holy days of the modern Druid calendar are the stations of the year, which track the movement of the Sun through the wheel of the seasons. Some other traditions take their holy days from the date on which a sacred scripture was revealed to the tradition's founder. The book of nature is the sacred scripture of Druidry, and it's always being revealed by the turning cycles of time and the changes of the natural world. The movements of the Sun set the great themes of the book of nature: birth, life, death, rebirth. The same sense of circling time that led the ancients to align Stonehenge on the midsummer sunrise still inspires Druids today.

Learning and Performing Rituals

Far too much confusion surrounds the subject of ritual. For many years, one of the great controversies in the Western world focused on whether ritual should be part of religious life. As is usual with binary thinking, the opposing sides went to extremes—one claiming that the exact performance of traditional rites was more important than anything else, the other insisting that ritual was by definition empty formalism or devil-worshipping magic. These two viewpoints—one overvaluing ritual and the other undervaluing it—still powerfully affect modern views on the subject and make clear thinking about ritual a good deal harder than it has to be.

Ross Nichols, one of the major figures in twentieth-century Druidry, offered a way around this confusion by proposing that "ritual is poetry in the world of acts." In this way of thinking, the performance of ritual has no special power of its own. Its power comes from the effects that it, like poetry, has on the people who take part in and experience it. Just as a well-written poem can reshape the awareness of its reader or hearer, revealing connections that might otherwise go unnoticed and highlighting neglected meanings, a well-performed ritual can do the same for those who take part in it.

Thus, celebrating the stations of the year isn't a mere formality. It focuses the experience of time and the seasons, lifting participants out of the limited consciousness of passing days into a wider awareness of the turning wheel of the year. It reminds them that their lives take place in a larger context, one in which other living beings and spiritual powers also have a place. It restores meaning to a world in which meaning often seems in short supply.

The mind also has unrecognized potentials that can be awakened through the redefining power of ritual. Thus ritual has effects that go well beyond the realm of psychology, and into subtle realms little understood by modern ways of approaching the world. For our purposes, it's enough to recognize that ritual is empty only when it's misused or ineptly performed. Done with skill and a grasp of basic principles, it can be full to bursting with meaning and can communicate that fullness to our everyday lives.

Whatever else can be said about it, ritual is a performing art related to drama, storytelling, and poetry recitation. What sets it apart from other performing arts is that the performers are also part of the audience. For this reason, the regular practice of ritual is the most important step in walking the Sun Path. By performing and experiencing rituals, novice ritualists learn from personal experience what works and what doesn't, and rehearse the skills required until they can do an effective ritual working at a moment's notice.

This sort of training yields the best results when the same ritual is practiced over and over, so that its structure and language become familiar and the practitioner can start paying closer attention to fine details and subtle factors. Performing a ritual you haven't practiced is like playing a piece of music you've never rehearsed: it takes a great deal of skill to do it well, and that level of skill comes within reach only when many other pieces have been practiced to the point of mastery.

In the next section, I give an opening and closing ritual for a Druid grove or ritual meeting designed to be performed and experienced by one person working alone. This is followed by rituals for the eight holy days of the modern Druid year, designed to fit between the ritual opening and the closing of the grove. These specific rituals aren't mandatory; you may choose instead to practice rituals from another Druid tradition, or to create your own. You may also choose to celebrate a different set of the Sun's stations. Although the two equinoxes and two solstices are fundamental in Druid Revival traditions and should not be neglected, you may choose not to celebrate the "cross-quarter days" between them, or decide to replace these with other celebrations that seem more relevant to your personal Druid path. The work required for the Sun Path is simply that you celebrate some set of holy days, including at least the solstices and equinoxes, during the year of your Druid initiation. If at all possible, take an active role in the celebrations; simply attending ceremonies as a spectator is a poor substitute for active engagement in the rituals of the seasons.

The ceremony below is a solitary ritual used to open and close the grove in the Ancient Order of Druids in America. The seasonal rituals following it are, on the other hand, original creations based on Druid Revival sources. In fact, there are no official AODA seasonal rituals. Indeed, writing and performing an original set of seasonal rituals is one of the requirements for AODA's second degree. The holy day rites given below are simply an example of what can be done with traditional materials, using methods of ritual and meditation common to most Druid paths.

Opening and Closing the Grove

The ceremony that follows is the basic framework for all the ritual work presented in this book, and can be used as a basis for many other kinds of ritual as well. The first half of the ceremony opens a circle of sacred space and brings that circle and the person inside it into contact with the powers of the Druid cosmos. When the work of the grove is done, the second half of the ceremony closes down the channels of power and brings the circle and the person inside it back into the world of ordinary experience.

If you choose to use the rituals in this book as the basis for your Sun Path work, plan on learning, memorizing, and practicing the grove ceremony relentlessly. A solid mastery of how to open and close ritual space and consciousness is essential to making your ritual work an effective and powerful experience. It's hard to concentrate on working with energy and entering the presence of the gods and goddesses if you have to fumble with a script or struggle to remember what you ought to be doing. Practice the grove ceremony as often as necessary to commit it to memory. Once a week during the year of your Druid initiation is a good target. If you work with a different set of rituals, of course, the same level of effort should be applied to them.

For the grove opening and closing ritual, find a space where you won't be disturbed. This can be indoors or out—although outdoors is traditionally preferred, indoors is sometimes more practical in today's world. There should be enough room for you to walk in a circle at least five or six feet across, and preferably twice that. In the center of the circle, place an altar. This can be a small table or stand with a flat top, or a convenient stone if you are outside; you can also simply spread an altar cloth on the ground, since the Earth itself is holy to Druids. For ordinary purposes, cover your altar with a white cloth, although colored cloths and other forms of decoration are often used in Druid seasonal ceremonies.

On the altar, place four small cauldrons or bowls, one on the northern, southern, eastern, and western sides of the altar. Fill the eastern cauldron half full of sand and place a stick or cone of incense in it. Fill the southern cauldron half full of sand as well and place a votive candle in it. Fill the western cauldron half full of pure water, preferably from a natural spring. Finally, fill the northern cauldron half full of salt or clean earth. These four cauldrons represent the four elements: air in the east, fire in the south, water in the west, and earth in the north. The incense, water, and salt or earth also represent the three Druid elements—nwyfre, gwyar, and calas—and the three rays of light with the candle as their source. Each of the seasonal rituals has additional requirements for the altar; you'll find these listed below.

On the northern edge of the circle, facing the altar, place a chair or some other object that marks it as a place to sit. If you're practicing outdoors, hauling a chair to your circle may be a nuisance; in this case, a convenient rock or simply a level place on the ground will certainly do. You will sit down here as the last act of the opening ceremony, and rise as the first act of the closing ceremony. The north is the position of honor in the grove, the place where the Chief Druid stands to face the Sun at its highest point in the sky. The altar is the Maen Log ("Stone of Speech") and the high place of light.

The ceremonies that follow call for three other items that can add much to the experience, but can also be dispensed with if necessary. The first is a sword with a sheath—preferably a straight double-edged sword of the sort used by knights in the Middle Ages. This represents Excalibur, the sword of King Arthur, one of the most evocative of all Druid symbols. In the opening ceremony, the sword is partly unsheathed and then returned as part of the formal proclamation of peace, while in the closing ceremony, the blade is fully drawn. Since you may not be able to use an actual sword in your ceremonies, I provide alternatives where necessary.

The second item is a drinking horn with a stand to hold it upright. This is filled with wine, mead, or apple cider, depending on your preferences and legal status, and represents the Hirlas, or horn of plenty, which has a central place in the seasonal rituals that follow. If you can't arrange for a drinking horn, a chalice or an ordinary glass will do, and you can substitute water for a stronger beverage in a pinch.

The third item is a white or unbleached robe, the traditional garment of Druids for the last 300 years. Some Druids wear robes with hoods, while others wear hoodless robes and nemysses—a simple cloth headdress of Egyptian style that came into use in the Druid movement around the beginning of the twentieth century (see figure 7). While wearing a robe is traditional, it's not required. You can perform Druid rituals just as well in some other sort of ritual garment; a famous picture of the nineteenth-century Welsh Druid Dr. William Price shows him bare-headed and barefoot, in what looks like long underwear of scarlet wool embroidered with Greek letters in bright green silk. Druids earlier in the Revival performed many rituals wearing ordinary clothing, with at most a colored ribbon tied around their upper arms to show their degree of initiation. Nudist Druids in the late nineteenth and twentieth centuries routinely performed their ceremonies with nothing on their bodies but sunlight and wind.

FIGURE 7. A DRUID NEMYSS.

Finally, each of the seasonal ceremonies below has an "offering and augury of plenty," a ritual gift traditionally given to all participants along with a drink from the Hirlas horn. The offering for each celebration is listed near the beginning of the text in the ceremonies that follow. Depending on the climate where you live, you may need to find substitutes for one or more of the traditional offerings; it won't do much good to go looking for snowdrops or other early spring flowers at the beginning of February, say, if they don't come up in your area for another three weeks. As always in Druidry, pay attention to your local environment and be willing to improvise. This will get you much further than a dour insistence on following rules.

OPENING THE GROVE

Prepare your ritual space by placing the altar cloth and any decorations on your altar and arranging the four cauldrons. If you have a sword, keep it sheathed and set it near the chair in the north. If you have a drinking horn, place it on its stand on the altar and fill it with whatever beverage you choose.

Light the incense and the lamp, and then go to the edge of the area where you'll be performing the ceremony. Take a few moments to clear your mind of unrelated thoughts and feelings. When you are ready, enter the circle and walk around it with the Sun—that is, clockwise—one full circle. Then walk around to the north side of the altar and face the south.

Raise your right palm forward to salute the Spiritual Sun, which is always symbolically at high noon in the southern sky, and say: *Let the powers attend as I am about to open a Grove of Druids in this place. The first duty of Druids assembled in the Sacred Grove is to proclaim peace to the four quarters of the world, for without peace our work cannot proceed.*

Take the sword from its place, still sheathed, and circle around to the east. Face outward and raise the sword in its sheath, holding it horizontally at head level, right hand on hilt, left on sheath.

Draw the sword partway from the sheath so that half the blade is visible, then push the sword back into the sheath. (If there's no sword available, raise your right palm outward to salute each direction instead.) Say: *I proclaim peace in the east.*

Lower the sword and proceed to the south, where you repeat the same process, drawing the sword partway and then sheathing it. Say: *I proclaim peace in the south.*

Proceed to the west, and do the same thing. Say: *I proclaim peace in the west.*

Proceed to the north and do the same thing. Say: *I proclaim peace in the north.*

Return the sword to its place and return to the north side of the altar, facing south across it. Say: *The four quarters are at peace and the work of the Grove may proceed. Let this Grove and all within it be purified with air.*

Go to the eastern side of the altar, pick up the cauldron with the incense, and carry it to the eastern edge of the space.

Walk in a clockwise circle once around the outer edge of the space, and concentrate on the idea that the incense smoke purifies and blesses everything in the grove. When you've come back around to the east, return the cauldron to the altar. Say: *Let this Grove and all within it be purified with fire.*

Take the cauldron with the lamp to the southern edge of the space and go once around clockwise, concentrating on the idea that the fire purifies and blesses everything in the grove. When you've come back around to the south, return the cauldron to the altar. Say: *Let this Grove and all within it be purified with water.*

Take the cauldron with the water to the western edge of the space and go once around clockwise, concentrating on the idea that the water purifies and blesses everything in the grove. When you've come back around to the west, return the cauldron to the altar. Say: *Let this Grove and all within it be purified with earth.*

Take the cauldron with the salt or earth to the northern edge of the space and go once around clockwise, concentrating on the idea that the salt or earth purifies and blesses everything in the grove. When you've come back around to the north, return the cauldron to the altar.

Standing at the north of the altar, say: *I invoke the blessing of the Mighty Ones with the words that have been the bond among all Druids:*

> Grant, O holy ones, thy protection;
> And in protection, strength;
> And in strength, understanding;
> And in understanding, knowledge;
> And in knowledge, the knowledge of justice;
> And in the knowledge of justice, the love of it;
> And in that love, the love of all existences;
> And in the love of all existences, the love of
> Earth our mother and all goodness.[9]

When you've finished the prayer, chant the word *Awen* three times. Draw the word out into its three syllables—Ah-Oh-En—and let it resonate throughout your body and the grove.

Go to the chair in the north and take your seat. This act completes the opening ceremony.

CLOSING THE GROVE

When the work of the grove is completed, sit down in the chair in the north and let your mind return to stillness.

When you're ready, rise and go to the north side of the altar, facing south across it. Say: *Let the powers attend as I am about to close a Grove of Druids in this place. Peace prevails in the four quarters and throughout the Grove. Let any power remaining from this working be returned to the Earth for its blessing.*

This is a crucial step, and takes practice. Any ritual working leaves some energy behind it, and this can usually be sensed as a mood, a feeling, or a subtle sense of presence in the space. Imagine this flowing inward toward the altar, down through it to the Earth, and then down to the Earth's center.

Keep concentrating on this until the ritual space feels clear of any leftover energy. Then say: *I now invoke the Sword of Swords.*

Draw the sword and hold it high, the blade pointing toward the sky. Say:

From the rising sun, three rays of light;

From the living earth, three stones of witness;

From the eye and mind and hand of wisdom,

Three rowan staves of all knowledge.

From the fire of the sun, the forge;

From the bones of the earth, the steel;

From the hand of the wise, the shaping;

From these, Excalibur.

By the Sword of Swords, I pledge my faithful service

To the living earth our home and mother.

If you don't have a sword, raise your right hand in salute to the Spiritual Sun in the south, and make the pledge in this way.)

Sheathe the sword and set it aside. Chant the word *Awen* three times, as in the opening ritual.

Raise your right hand in salute to the Spiritual Sun in the south, leave the altar, and walk in a clockwise circle around the space. Continue around to the exit and leave the grove. This concludes the closing ceremony.

Holy Day Rituals

Celebrating the holy days of the Druid calendar is the most important use you'll make of the skills and techniques of ritual performance. Self-initiation into the Druid tradition, however important its effects may be, is a process you'll pass through only once, but the wheel of the year will carry you from season to season as long as you remain a Druid. There's a sense in which the stations of the year are your real initiation, repeated at ever deeper levels as one year leads to the next.

The rituals given here are designed for the temperate regions of the Northern Hemisphere. They call on gods and goddesses central to many British Druid Revival traditions. If you live in the Southern Hemisphere or in a place with significantly different seasons, or if you prefer to invoke different spiritual powers, you can modify the texts below accordingly. This is entirely appropriate; it makes no sense to perform a midsummer ritual at the Southern Hemisphere's midwinter just because a book tells you to do it in June. Likewise, it makes no sense to attune yourself with some other place's seasonal rhythms rather than your own, or to call on gods you don't know and neglect the ones you do. The following rituals can therefore be used as is or modified to fit your local ecology and personal theology. The dates can also be varied, within reason, to fit the demands of your schedule. Many Druid groves, for example, perform their rituals on the Saturday or Sunday closest to the actual date of the holy day in question.

The rituals require some familiarity with the basic skills of meditation, including color breathing and discursive meditation. If you plan to use these ceremonies during the year of your Druid initiation, take the time to familiarize yourself with these skills before your first holy day ritual (see chapter 9).

Each ritual has an "offering and augury of plenty" that is presented to the gods in the course of the ceremony. One part of this is the Hirlas, the horn of plenty, which contains wine, mead, or cider. The other part varies with the seasons. Edible offerings, such as bread and fruit, should be eaten with reverence after the ceremony. The inedible offerings should be put in a special place, such as a personal altar. They can be kept until the next holy day and then returned to the Earth in some suitable way.

SAMHUINN

Celebrated on or near 1 November, Samhuinn was the beginning of the traditional Celtic year, yet its name—spelled Samhain in modern Irish—means "summer's end." This mirrors the central Druid tradition that every end is a beginning and every beginning an end, in a circle of completion and renewal. In this spirit, the festival of Samhuinn honored the dead in old times, and still plays that role in traditional Christianity (as All Saints' Day and All Souls' Day). This is also true in many other cultures. Tradition holds that the veil between the worlds of the living and the dead is thin at this time. Since death, in Druid teaching, is a transformation rather than an end, the evergreen, the emblem of enduring life, is the chief symbol of this season of the year.

Ceremony:

Before beginning the ritual, place a sprig of evergreen on the altar along with the Hirlas horn. The altar may be draped with fall colors such as harvest gold, russet, and brown, or decorated with fall leaves; use your imagination and your creative abilities. Once all is ready, open the grove with the ceremony given above, take your seat in the north, and enter into meditation with a cleansing breath. Use violet color breathing to prepare for the meditation.

The theme for meditation, which should be read aloud in a quiet voice, is as follows:

> This day of Samhuinn in the Druid tradition marks the ending of one year and the beginning of another. In this season, the last of the harvest is brought from the fields. The Sun, our father, descends into darkness, and the Earth, our mother, dons her brown garments of mourning. The veil between the worlds becomes thin, and the ancestors come close to us. Their voices whisper in the autumn winds. As we stand among the falling leaves and the gray and golden light, let us remember the past and its lessons, and gather a harvest of wisdom to bear us through winter to the new spring to come.

Meditate on this for a time, then finish the meditation and rise to your feet, facing the altar. Say aloud:

> On this day I invoke Ceridwen the goddess of wisdom, keeper of the cauldron of transformation, mistress of the cycles of change. Ceridwen. Ceridwen. Ceridwen. Join with me in the circle of the Grove. Grant me your blessings, and receive my blessing in return. Watch over me and over the living world; protect, cherish and guide all existences in the turning of the year's wheel now before me.

Concentrate on the presence of the goddess. When you can feel her in the grove, go to the altar and take up the Hirlas horn and the evergreen sprig. Say:

> In the presence of the goddess, I bear the offering and augury of plenty. From the green realms of the Earth Mother, the spirits of nature gather gifts for our sustenance, each in its own place in the wheel of the turning year. On this day of Samhain, behold the horn of plenty, emblem of the abundant gifts of nature, and the sprig of evergreen, in recollection of the year that is past and hope for the year now to come. In these symbolic gifts, may there be blessing on the Earth forever.

Raise the horn up high, lower it again, and drink from it. Then place the horn on the altar and say:

> From the Earth, her never failing promise; from the Holy Kindreds, their gifts of will and grace. I receive with thanks the evergreen of enduring life. From all that is given, I give in turn.

Lay the sprig upon the altar in offering, return to the chair and resume your seat. Say:

> Those who have established the turnings of the silent stars, who place the seal of their blessing on the living Earth and lift up its children to inscrutable heights, have given to all without tribute and without price. May I be worthy of that which is given.

Pause and say:

> With thanks, I have received the blessing of Ceridwen.

Sense the departure of the goddess. Then close the grove with the ceremony given above.

ALBAN ARTHUAN

Alban Arthuan, or Alban Arthan in modern Welsh, is the first of the four Alban Gates or solar festivals in the year. It is celebrated on or around 21 December, the time of the winter solstice. The name traditionally means "The Light of Arthur." The solstice, the day when the Sun finishes its long retreat into the southern sky and begins to move northward once again, was seen as the birthday of the new Sun, or of the hero who symbolized the Sun in many old mythologies. Many modern Druids transfer much of the symbolism of the Christian festival of Christmas to this holy day. Mistletoe is this day's special emblem. Alban Arthuan is a time for firelight, feasting, and rejoicing as the new Sun returns with the promise of summer to come.

Ceremony:

Before you begin, place a sprig of mistletoe on the altar along with the Hirlas horn. The altar may be decorated in green and white. Open the grove with the usual ceremony, take your seat in the north, and enter into meditation with a cleansing breath. The color breathing should be white.

The theme for meditation, which should be read aloud in a quiet voice, is as follows:

> This turning of the year wheel marks the shortest day and greatest night, the lower hinge of the year, when the Sun completes his descent into darkness and begins his return. The Mabon is born in darkness, small and still; the star is lit in the heart of the rose, the new fire returns in the place of light, the first ray of the reborn Sun descends into the Earth's womb in the great mound of Aengus. It was on this day, likewise, that Arthur began and fulfilled the circle of his destiny across the twelve sieges of the turning table of the stars, and on this day shall he begin it anew in time to come. It is for this reason that this day is called Alban Arthuan, the Light of Arthur. From the point of light in the midst of darkness unfolds the creation of all things, the revolution of all that is past toward a new good.

Meditate on this for a time, then finish the meditation and rise to your feet, facing the altar. Say aloud:

> On this day, I invoke Hu Gadarn, Hu the Mighty, the infinitely great and infinitely small, master of the shining stars and the great deep, high god of the Druid mysteries. Hu. Hu. Hu. Join with me in the circle of the Grove. Grant me your blessings, and receive my blessing in return. Watch over me and over the living world; protect, cherish, and guide all existences in the turning of the year's wheel now before me.

Concentrate on the presence of the god. When you can feel him in the grove, go to the altar and take up the Hirlas horn and the mistletoe sprig. Say:

> In the presence of the god, I bear the offering and augury of plenty. From the green realms of the Earth Mother, the spirits of nature gather gifts for our sustenance, each in its own place in the wheel of the turning year. On this day of Alban Arthuan, behold the horn of plenty, emblem of the abundant gifts of nature, and the mistletoe, symbol of the presence of divinity in all things. In these symbolic gifts, may there be blessing on the Earth forever.

Raise the horn up high, lower it again, and drink from it. Then place the horn on the altar and say:

> From the Earth, her never failing promise; from the Holy Kindreds, their gifts of will and grace. I receive with thanks the mistletoe that represents the ever-renewed blessing of the gods and goddesses. From all that is given I give in turn.

Lay the sprig on the altar as an offering, return to the chair, and resume your seat. Say:

> Those who have established the turnings of the silent stars, who place the seal of their blessing on the living Earth and lift up its children to inscrutable heights, have given to all without tribute and without price. May I be worthy of that which is given.

Pause and say:

> With thanks, I have received the blessing of Hu.

Sense the departure of the god. Then close the grove with the usual ceremony.

IMBOLC

This festival comes on or around 2 February. Its name—also spelled Oimelc in some sources—literally means "ewe's milk," since it comes at the beginning of the lambing season, one of the first signs of returning life in the cold, gray days of winter. It is also the time of the first flowers in much of the temperate world, and early flowers such as the snowdrop are its chief symbol. In many Druid traditions, this is a goddess-centered ritual of light and water, a time of purification and renewal focused on the stirrings of life in the womb of the Earth.

Ceremony:

Before you begin, place a snowdrop or other early spring flower on the altar along with the Hirlas horn. The altar may be decorated in blue and silver. Once all is ready, open the grove with the usual ceremony, take your seat in the north, and enter into meditation with a cleansing breath. The color breathing should be indigo.

The theme for meditation, which should be read aloud in a quiet voice, is as follows:

> We have come to the time of the year's quickening, the first stirrings of spring in the womb of the Earth Mother. Fire is now manifested as light, the rising light of spring emerging from the creative waters, the horizontal plane of space bisected by the vertical rays of time. Now is the time of the first plow, the birth of lambs in the pastures, the washing of the face of the Earth, and the blessing of candles. The torches burn as the young goddess returns to the waxing day; the snowdrops bloom amid the melting snows. On this day of Imbolc are renewed once again the bonds that are made between heaven and Earth.

Meditate on this for a time, then finish the meditation and rise to your feet, facing the altar. Say aloud:

> On this day, I invoke Ana, the mother of gods and goddesses, the mother of all things, great goddess of the cosmos in whose presence and by whose bounty I exist. Ana. Ana. Ana. Join with me in the circle of the Grove. Grant me your blessings, and receive my blessing in return. Watch over me and over the living world; protect, cherish, and guide all existences in the turning of the year's wheel now before me.

Concentrate on the presence of the goddess. When you can feel her in the grove, go to the altar and take up the Hirlas horn and the snow-drop. Say:

> In the presence of the goddess, I bear the offering and augury of plenty. From the green realms of the Earth Mother, the spirits of nature gather gifts for our sustenance, each in its own place in the wheel of the turning year. On this day of Imbolc, behold the horn of plenty, emblem of the abundant gifts of nature, and the snowdrop [or whatever flower you are using], the first flower of spring, representing the waking of light and life in the depths of the winter's darkness. In these symbolic gifts, may there be blessing on the Earth forever.

Raise the horn up high, lower it again, and drink from it. Then place the horn on the altar and say:

> From the Earth, her never failing promise; from the Holy Kindreds, their gifts of will and grace. I receive with thanks the first flower of the returning spring. From all that is given, I give in turn.

Lay the sprig upon the altar in offering, return to the chair and resume your seat. Say:

> Those who have established the turnings of the silent stars, who place the seal of their blessing on the living Earth and lift up its children to inscrutable heights, have given to all without tribute and without price. May I be worthy of that which is given.

Pause and say:

> With thanks, I have received the blessing of Ana.

Sense the departure of the goddess. Then close the grove with the usual ceremony.

ALBAN EILER

The second of the Alban Gates, Alban Eiler is celebrated on or around 21 March, at the time of the spring equinox. The name Alban Eilir in modern Welsh is interpreted as "Light of the Earth" in Druid tradition. This is the season when the energies of fertility are at their height. Alchemists once went out at first light during the time from Alban Eiler to Belteinne to gather dew from the wet grass, knowing that, at this time of year, it was charged with powerful energies of renewal and life.

Ceremony:

Before you begin, put a vase of spring flowers and a bowl containing some flower or vegetable seeds on the altar along with the Hirlas horn. The altar may be decorated in shades of green. Once all is ready, open the grove with the usual ceremony, take your seat in the north, and enter into meditation with a cleansing breath. The color breathing should be red.

The theme for meditation, which should be read aloud in a quiet voice, is as follows:

> In the turning wheel of the year, now is the moment of equal day and night, the time when the promise of Earth moves toward its summer fulfillment. This is the time of Alban Eiler, the Light of the Earth. The birds return from southern lands bearing springtime beneath their wings. The flowers burgeon with the promise of fruit. The green blades of the grainfields leap toward the Sun. From all that is given us, we build again the temple and song of the Alban Gate.

Meditate on this for a time, then finish the meditation and rise to your feet, facing the altar. Say aloud:

> On this day, I invoke Coel, mighty and hidden spirit of life, master of the fires of spring and the leap of stag and salmon, power of vitality surging through the Earth's veins. Coel. Coel. Coel. Join with me in the circle of the Grove. Grant me your bless-

ings, and receive my blessing in return. Watch over me and over the living world; protect, cherish, and guide all existences in the turning of the year's wheel now before me.

Concentrate on the presence of the god. When you can feel him in the grove, go to the altar and take up the Hirlas horn, the flowers, and the seeds. Say:

In the presence of the goddess, I bear the offering and augury of plenty. From the green realms of the Earth Mother, the spirits of nature gather gifts for our sustenance, each in its own place in the wheel of the turning year. On this day of Alban Eiler, behold the horn of plenty, emblem of the abundant gifts of nature; spring flowers, the blessing of life and beauty that dances now upon the land, and seeds for the gardens of Earth, in earnest of greater blessings to come. In these symbolic gifts, may there be blessing on the Earth forever.

Raise the horn up high, lower it again, and drink from it. Then place the horn on the altar and say:

From the Earth, her never failing promise; from the Holy Kindreds, their gifts of will and grace. I receive with thanks the flowers of beauty and the seeds of new life. From all that is given, I give in turn.

Lay the flowers and seeds upon the altar in offering, return to the chair and resume your seat. Say:

Those who have established the turnings of the silent stars, who place the seal of their blessing on the living Earth and lift up its children to inscrutable heights, have given to all without tribute and without price. May I be worthy of that which is given.

Pause and say:

With thanks, I have received the blessing of Coel.

Sense the departure of the god. Then close the grove with the usual ceremony.

BELTEINNE

At the opposite pole of the turning year from Samhuinn, on or around 1 May, is Belteinne, the great Druid feast of life and love, celebrating the triumph of spring and the renewal of the Earth. This is symbolically the day when the Sun weds the Earth and impregnates her, engendering the new life that will stir again nine months later at Imbolc of the next year. Hawthorn is the great emblem of this festival and rejoicing its major theme.

Ceremony:

Before you begin, put a sprig of hawthorn on the altar along with the Hirlas horn. The altar may be decorated in white. Once all is ready, open the grove with the usual ceremony, take your seat in the north, and enter into meditation with a cleansing breath. The color breathing should be orange.

The theme for meditation, which should be read aloud in a quiet voice, is as follows:

> On this day of Belteinne, spring stands before us in her fullness. It is the season of flowers, and the Earth welcomes the embrace of the Sun. On this day in ancient time, fire blazed upon the hill, symbol of the renewal of light, while offerings were made at well and spring in the valley below. Fire and water, holy hill and holy well, stand as twin pillars of the gateway, and between them is the third, in which they become one—the Tree that spans the worlds from height to depth.

Meditate on this for a time, then finish the meditation and rise to your feet, facing the altar. Say aloud:

> On this day, I invoke Niwalen, the young goddess of the waking year, laughing among the flowers, guardian of the sacred spring and the ancient track. Niwalen. Niwalen. Niwalen. Join with me in the circle of the Grove. Grant me your blessings, and receive my blessing in return. Watch over me and over the living world; protect, cherish, and guide all existences in the turning of the year's wheel now before me.

Concentrate on the presence of the goddess. When you can feel her in the grove, go to the altar and take up the Hirlas and the hawthorn sprig. Say:

> In the presence of the goddess, I bear the offering and augury of plenty. From the green realms of the Earth Mother, the spirits of nature gather gifts for our sustenance, each in its own place in the wheel of the turning year. On this day of Belteinne, behold the horn of plenty, emblem of the abundant gifts of nature, and sprigs of hawthorn, the symbol of blessing and of purity. In these symbolic gifts, may there be blessing on the Earth forever.

Raise the horn up high, lower it again, and drink from it. Then place the horn on the altar and say:

> From the Earth, her never failing promise; from the Holy Kindreds, their gifts of will and grace. I receive with thanks the hawthorn sign of purity. From all that is given, I give in turn.

Lay the sprig upon the altar in offering, return to the chair, and resume your seat. Say:

> Those who have established the turnings of the silent stars, who place the seal of their blessing on the living Earth and lift up its children to inscrutable heights, have given to all without tribute and without price. May I be worthy of that which is given.

Pause and say:

> With thanks, I have received the blessing of Niwalen.

Sense the departure of the goddess. Then close the grove with the usual ceremony.

ALBAN HERUIN

At Alban Heruin, the third of the Alban Gates, the Sun is at his highest point in the heavens, but the triumph of light is shadowed by the recognition that afterward comes the gradual descent into winter's darkness. This festival, called Alban Hefin in modern Welsh, is celebrated on or about 21 June, the date of the summer solstice. Its name is traditionally interpreted as "the Light of the Shore," the point of contact between the Earth (Alban Eiler, the Light of the Earth) and the sea (Alban Elued, the Light of the Sea). Vervain, the holy healing plant of the ancient Druids, is a special emblem of this day.

Ceremony:

Before you begin, place a sprig of vervain on the altar along with the Hirlas horn. The altar may be decorated in red and gold. Once all is ready, open the grove with the usual ceremony, take your seat in the north, and enter into meditation with a cleansing breath. The color breathing should be golden yellow.

The theme for meditation, which should be read aloud in a quiet voice, is as follows:

> This day of Alban Heruin is the hinge and center of the year. The Sun stands at meridian height. At this time, the bond between heaven and Earth is strongest, and the Sun, the regent of light and the center of the spinning worlds, makes most urgent his message at this turning point of the year. The meaning of Alban Heruin is the Light of the Shore, for the shore is the meeting place of sea and land, where the two borders tremble together in a circle around the Earth. Thus we celebrate the glory of high summer, remembering the winter past and the winter to come.

Meditate on this for a time, then finish the meditation and rise to your feet, facing the altar. Say aloud:

> On this day, I invoke Beli, the strong god of the Sun's zenith, lord of the threefold fire. Beli. Beli. Beli. Join with me in the circle of the Grove. Grant me your blessings, and receive my blessing in return. Watch over me and over the living world; protect, cherish, and guide all existences in the turning of the year's wheel now before me.

Concentrate on the presence of the god. When you can feel him in the grove, go to the altar and take up the Hirlas and the vervain sprig. Say:

> In the presence of the god, I bear the offering and augury of plenty. From the green realms of the Earth Mother, the spirits of nature gather gifts for our sustenance, each in its own place in

the wheel of the turning year. On this day of Alban Heruin, behold the horn of plenty, emblem of the abundant gifts of nature, and vervain, the healing herb of the ancient Druids and the token of sight and insight. In these symbolic gifts, may there be blessing on the Earth forever.

Raise the horn up high, lower it again, and drink from it. Then place the horn on the altar and say:

From the Earth, her never failing promise; from the Holy Kindreds, their gifts of will and grace. I receive with thanks the blessing of vervain, the emblem of healing and of vision. From all that is given I give in turn.

Lay the sprig upon the altar in offering, return to the chair and resume your seat. Say:

Those who have established the turnings of the silent stars, who place the seal of their blessing on the living Earth and lift up its children to inscrutable heights, have given to all without tribute and without price. May I be worthy of that which is given.

Pause and say:

With thanks, I have received the blessing of Beli.

Sense the departure of the god. Then close the grove with the usual ceremony.

LUGHNASADH

At the beginning of the harvest, on or around 1 August, Lughnasadh marks the first of the three festivals of the declining light. Its name means "festival of Lugh." Irish myths relate it to Lugh, the Sun god, who established it in honor of his foster-mother Tailtiu, the goddess of agriculture. Bread made from newly harvested grain is a central symbol of this holy day, and many modern Druids celebrate this—and indeed all three of the autumn ceremonies—with a feast following the ceremony.

Ceremony:

Before you begin, place a piece of freshly baked bread on the altar along with the Hirlas horn. Any kind of bread you prefer and enjoy baking can be used; barley bread is traditional in certain Druid orders, while some Druids in North America consider corn (maize) bread best suited to their harvest celebrations. The altar may be decorated in green and gold. Once all is ready, open the grove with the usual ceremony, take your seat in the north, and enter into meditation with a cleansing breath. The color breathing should be green.

The theme for meditation, which should be read aloud in a quiet voice, is as follows:

> This day of Lughnasadh marks the beginning of the Sun's decline from the high places of heaven, and the beginning of the harvest upon the breast of the Earth. The year wheel from this season rolls downward into the deep places of winter and the rekindling of new light. The Sun descends in the sky, into autumn, but with glory. The wheel is always balanced, and what seems to be lost or declining is regained elsewhere or at another time, for rebirth is the perpetual law of nature. Thus we gather the fruits of the marriage of Sun and Earth.

Meditate on this for a time, then finish the meditation and rise to your feet, facing the altar. Say aloud:

> On this day, I invoke Sul, the healing goddess of the bright Sun and the blessed spring, mistress of the fire above and the fire below. Sul. Sul. Sul. Join with me in the circle of the Grove. Grant me your blessings, and receive my blessing in return. Watch over me and over the living world; protect, cherish, and guide all existences in the turning of the year's wheel now before me.

Concentrate on the presence of the goddess. When you can feel her in the grove, go to the altar and take up the Hirlas and the bread. Say:

> In the presence of the goddess, I bear the offering and augury of plenty. From the green realms of the Earth Mother, the spirits of nature gather gifts for our sustenance, each in its own place in the wheel of the turning year. On this day of Lughnasadh,

> behold the horn of plenty, emblem of the abundant gifts of nature, and bread freshly baked, in commemoration of the harvest now beginning. In these symbolic gifts, may there be blessing on the Earth forever.

Raise the horn up high, lower it again, and drink from it. Then place the horn on the altar and say:

> From the Earth, her never failing promise; from the Holy Kindreds, their gifts of will and grace. I receive with thanks the bread of the new harvest. From all that is given, I give in turn.

Lay the bread upon the altar in offering, return to the chair in the west, and resume your seat. Say:

> Those who have established the turnings of the silent stars, who place the seal of their blessing on the living Earth and lift up its children to inscrutable heights, have given to all without tribute and without price. May I be worthy of that which is given.

Pause and say:

> With thanks, I have received the blessing of Sul.

Sense the departure of the goddess. Then close the Grove with the usual ceremony.

ALBAN ELUED

On or around the day of the fall equinox, 23 September, Alban Elued marks the midpoint of the Sun's decline and the heart of the harvest season. Its name, Alban Elfed in modern Welsh, is interpreted in Druid tradition as "the Light of the Sea," representing the descent of the Sun into the western seas toward the close of the day and the year. As the fourth of the Alban Gates, it marks the outflowing of the energy that poured into the Earth with the spring. Its chief emblem is fruit, heralding the fruits of the Earth that attain ripeness around this time.

Ceremony:

Before you begin, set a piece of fruit on the altar along with the Hirlas horn. The altar may be decorated in green and blue. Once all is ready, open the grove with the usual ceremony, take your seat in the north, and enter into meditation with a cleansing breath. The color breathing should be sky blue.

The theme for meditation, which should be read aloud in a quiet voice, is as follows:

> With the coming of Alban Elued, the Light of the Sea, the wheel of the turning year rolls toward its end and beginning. It is the time of recollection and return. Now the leaves turn from green to gold and descend with the waning Sun into the depths of the Earth Mother. Now the sea becomes dark as the year descends into ocean. In silence, the seed of wisdom is gathered up, to be stored through the winter that it may become new life in the spring. From all that is given us, we build again the temple and song of the Alban Gate.

Meditate on this for a time, then finish the meditation and rise to your feet, facing the altar. Say aloud:

> On this day, I invoke Esus, chief of all tree-spirits, who sits in the first fork of the sacred oak. Esus. Esus. Esus. Join with me in the circle of the Grove. Grant me your blessings, and receive my blessing in return. Watch over me and over the living world; protect, cherish, and guide all existences in the turning of the year's wheel now before me.

Concentrate on the presence of the god. When you can feel him in the grove, go to the altar and lift up the Hirlas horn and the fruit. Say:

> In the presence of the god, I bear the offering and augury of plenty. From the green realms of the Earth Mother, the spirits of nature gather gifts for our sustenance, each in its own place in the wheel of the turning year. On this day of Alban Elued, behold the horn of plenty, emblem of the abundant gifts of nature, and fruit ripe from the summer's warmth and the cooling nights of autumn, the richness of the harvest and the gifts of the Earth Mother. In these symbolic gifts, may there be blessing on the Earth forever.

Raise the horn up high, lower it again, and drink from it. Then place the horn on the altar and say:

> From the Earth, her never failing promise; from the Holy
> Kindreds, their gifts of will and grace. We receive with thanks
> the fruit of the harvest. From all that is given, we give in turn.

Lay the fruit upon the altar in offering, return to the chair, and resume your seat. Say:

> Those who have established the turnings of the silent stars, who
> place the seal of their blessing on the living Earth and lift up its
> children to inscrutable heights, have given to all without tribute
> and without price. May I be worthy of that which is given.

Pause and say:

> With thanks, I have received the blessing of Esus.

Sense the departure of the god. Then close the grove with the usual ceremony.

Further Reading

Fortunately for the modern Druid, there is a substantial literature on seasonal traditions in British folk culture and modern Druid practice. The following books are worth reading as sources for your own holy day celebrations:

Janet and Colin Bord, *Earth Rites* (London: Granada, 1982). A good general summary of British seasonal celebrations, with much information on traditions surrounding the eightfold year.

Marian Green, *A Harvest of Festivals* (London: Longman, 1980). Another good introduction to the seasonal festivals and traditions of Britain.

Ellen Evert Hopman, *A Druid's Herbal for the Sacred Earth Year* (Rochester, VT: Destiny, 1995). An elegant introduction to herblore arranged according to the eight holy days of the modern Druid calendar.

Ronald Hutton, *The Stations of the Sun* (Oxford: Oxford University Press, 1996). A scholarly study of seasonal celebrations in Britain, focusing on the ways in which folk traditions have changed creatively over time.

John King, *The Celtic Druid's Year* (London: Blandford, 1994). A useful survey of the calendar systems associated with ancient and modern Druids.

Alexei Kondratiev, *The Apple Branch: A Path to Celtic Ritual* (Cork: Collins Press, 1998). A practical guide to seasonal ceremonies and other ritual work from the standpoint of Celtic Reconstructionist spirituality, well worth reading.

Maire MacNeill, *The Festival of Lughnasa* (Oxford: Oxford University Press, 1962). A classic work of scholarship exploring the traditional Irish celebration of Lughnasadh in ancient and modern times.

Margie McArthur, *Wisdom of the Elements* (Freedom, CA: Crossing Press, 1998). A magical workbook based on the four elements of medieval philosophy and magic, this book has much to offer to ritual designers working with the wheel of the year.

Claire Nahmad, *Earth Magic* (Rochester, VT: Destiny, 1994). A journey through the cycle of the year enriched with herbal knowledge, folk magic, weatherlore, and astrological traditions.

Nigel Pennick, *Practical Magic in the Northern Tradition* (Wellingborough, Northants: Aquarian, 1989). The best attempt so far to recreate the worldview and cosmology of traditional northwestern European magical practice, with much material on seasonal cycles and celebrations.

9

THE MOON PATH

THE THIRD PART OF THE TRIAD
that forms the core of Druid initiation and practice is the
Moon Path, the way of meditation. The Moon reflects the
light of the Sun, illuminating the night sky of Earth. Less
obviously, it also reflects the blue radiance of the Earth,
casting back a pale glow that can be seen on a clear night
when the Moon is a thin crescent.

In much the same way, the human mind reflects the
brightly colored world around us, and it also more subtly
reflects hidden worlds within us. Just as the Moon mediates light between Sun and Earth, the mind mediates
awareness between matter and spirit. In fact, the words
"moon" and "mind" come from the same ancient word for
"measure."

In many ways, the Moon Path of meditation is the heart and soul of the Druid way. Without it, the Earth and Sun Paths can turn into outward formalities lacking any spiritual dimension. The gifts brought by meditation give these paths their inner life, making both of them vehicles for the opening of a wider awareness of the universe and the self.

Some people involved in the more political end of the environmental movement have assailed inner development as an evasion of responsibility, an "awareness trap" that draws energy away from the world of outward action. While this sort of claim is understandable, it's also shortsighted. In his powerful study of the crisis of modern industrial culture, *Where the Wasteland Ends*, Theodore Roszak points out that the damage we have done to the living Earth is a reflection of the damage we have done to ourselves:

> What, after all, is the ecological crisis that now captures so much belated attention but the inevitable extroversion of a blighted psyche? Like inside, like outside. In the eleventh hour, the very physical environment suddenly looms up before us as the outward mirror of our inner condition, for many the first discernible symptom of advanced disease within.[10]

Thus the quest to restore a sane relationship between human society and the natural world demands more from us than outward changes in our lives. It requires new awareness and knowledge. In Druidry, these are sought and found through meditation. In the same way, the Moon Path both reflects and enhances the heightened awareness of natural cycles and spiritual powers that comes through the Sun Path, the ritual celebration of the turning year.

Nobody knows what practices the ancient Druids taught to their students, but it's a good bet that some form of meditation was among them. Nearly all the world's spiritual traditions give meditation a central place among methods of inner development. Certainly, the Druid Revival took up meditation as a central practice early on. History had a large part in this, for the art of meditation was going through a major rebirth of its own in England when the Druid Revival emerged. All through the seventeenth and eighteenth centuries, the Church of England made meditation an important part of its teachings. A whole literature of meditation handbooks, mostly forgotten now, gave Anglican clergy and churchgoers alike the tools to add an inner dimension to the formal side of their religion. The connections between early Druidry and the Anglican Church gave Druids ready access to all this material.

Later on, as Druidry moved away from its origins and pursued its own destiny, new sources of meditation techniques came into the tradition. Many nineteenth-century Druids avidly read English translations of Hindu mystical and spiritual texts. Later, the overlap between Druidry and groups such as the Theosophical Society and the Hermetic Order of the Golden Dawn brought more methods of meditation into Druid tradition. By the dawn of the twenty-first century, meditation had been central to the Druid tradition for well over 200 years.

There are many ways to meditate, and each has its own strengths and weaknesses. The way given here is no exception. It has two chief strengths. First, it doesn't need personal teaching or supervision; it can be learned effectively and practiced safely out of a book. Second, it doesn't require major lifestyle changes or inborn talent; anyone in any walk of life willing to set aside fifteen minutes a day can practice it and get results. Its drawbacks, in turn, are the other side of its strengths: because it's a gentle, low-intensity approach to meditation, it takes time, and its effects unfold gradually and subtly. Certain intensive methods of meditation produce spectacular results in short order, but they involve equally spectacular risks to mental and physical health. Don't expect either of those here.

If you've practiced Asian systems of meditation, one feature of Druid meditation may surprise you. Many Eastern methods treat ordinary thinking as an obstacle to be overcome. "The mind is the slayer of the Real; slay thou the slayer," is a maxim sometimes repeated in these circles. From a Druid perspective, this approach has problems. Methods based on stopping the mind work, but they too often produce people who are spiritually enlightened but intellectually inept. This has practical drawbacks, and it also makes a poor fit with Druidry, which respects knowledge as well as spiritual insight. The Druid way of meditation, like most other Western meditative systems, takes a different approach. It trains and directs thought instead of silencing it.

The core process of meditation, no matter what system you're discussing, is focused attention. The Druid meditator focuses attention on thinking itself, directing thought steadily onto a chosen subject. The mind thus unfolds its inner potential, as in any other kind of meditation, and the meditator also comes to understand more about the subject of the meditation.

This form of meditation is called discursive meditation, because it often takes the form of an inner discourse or dialogue. It can be defined as the continued movement of thought toward a chosen theme. To practice it, arrange an appropriate place, time, and physical state, and take a posture that aids meditation. Relax into that posture and use patterns of breathing to help bring your physical body and nwyfre into quiet balance. A short period of concentration stills the ordinary chatter of thought, and then your mind turns to the subject of meditation. Keep your thoughts focused on the subject, or bring them back if they wander away. As you call to mind each aspect of the subject and follow out its implications, you learn to shape the flow of thought, while unfolding hidden dimensions of the subject itself.

This process requires some learning, of course. Most novice meditators find it easiest to start with practical details such as place, time, and posture, and proceed inward from there, through a series of exer-

cises in which each step includes all the previous steps. We'll begin
with preliminaries on the physical level.

Preliminaries to Meditation

You'll need a private place for practice with no loud noise sources
nearby, and with a clock you can see easily. The place doesn't have to
be particularly fancy or Druidical. A room with a door you can close
is all that's required. If you can't manage this, a corner of a room and
a little forbearance on the part of your housemates will do the job.
Many Druids like to meditate outdoors, and this can work well if your
local climate and human noises don't interfere. Meditating a few yards
from a waterfall is one thing; meditating a few yards from a dirt road
used by motorcycles is quite another.

In the early stages of your training, it's a good idea always to meditate in the same place. The mind is a creature of habit, and if you
always go to the same place when you meditate, it will be ready to
enter a meditative state whenever you do so.

Clothing for meditation should be comfortable and loose, especially
around the waist and chest. Tight belts and tight bras are particularly
unhelpful. Avoid anything that restricts breathing or puts pressure on
your body. Many Druids find that a robe works well for meditation; if
you use a robe for ritual, you can certainly wear it for meditation as
well.

Time is also a factor. Making time for daily meditation is the most
important step you'll take on the Moon Path. It's also the step that
beginners resist most stubbornly, even though it usually takes nothing
more than setting the alarm clock a little earlier in the morning or
skipping one TV sitcom each evening. Many students start out medi-

tating once or twice a week. This is fine to begin with, as long as you stick to it. Don't let your meditation slide into the heap of things you want to get back to someday. Occasional meditation is better than none, but the benefits of meditation only come with regular practice.

Most students find it best to meditate at the same time every day, to make the best use of the power of habit. This can be either the same clock time or, if your schedule swings around the clock, the same point in your daily activities—for example, in the morning before breakfast or just before going to bed at night. If you live with other people, of course, your meditation time will need to be negotiated with them.

In today's reality of busy schedules and crowded evenings, it's often best to make time for meditation by setting the alarm half an hour early each morning. Meditating first thing in the morning allows you to come to your meditation with a mind as uncluttered as possible, and it makes sure that your meditation won't be pushed aside if the time demands of your job or social life spin out of control. It also makes the most use of meditative themes as sleep subjects, a technique explained later on.

Your practices and your lifestyle will impact each other, although much less drastically than some traditions teach. You don't have to take up a special diet and retreat to a monastery in order to practice meditation. However, food and drink do affect your ability to meditate. It's best not to meditate within about an hour after eating a meal; the heavier the meal, the longer you should wait. Digestion sends most of the body's resources to the stomach, while meditation sends most of them to the brain. Trying to do both at once can produce indigestion and nausea, as well as muddled meditations.

Some traditions insist that meditators take up a vegetarian diet. What you eat affects your awareness, and this can be a tool worth using in certain situations. Still, insisting that any spiritually minded person must follow a particular diet is both inaccurate and potentially harmful. It's inaccurate because the world contains vibrant meditative traditions that counsel every diet imaginable. Meat, onions, salt, grain products, and just about every other food you can name are forbidden by some traditions, mandated by others, and treated as unimportant by still others. (People who think that eating meat and having powerful spiritual experiences are mutually exclusive, for instance, should sit down with a copy of John Neihardt's book *Black Elk Speaks* and a *very* strong cup of chamomile tea.) It's potentially harmful because people have different nutritional needs, and a diet that feeds one person perfectly can leave another malnourished. However spiritual your intentions may be, if what you're doing damages your health, you're doing something wrong.

The case of sex is similar. You'll hear claims that meditation should only be performed by people willing to take and keep a vow of chastity. This is true of some intense and dangerous methods that redirect sexual energies into other channels. It's not even remotely true of the kind of meditation taught here. Different people have different responses to sex; some find it best to wait for an hour or so after intercourse to allow subtle energies to return to normal, while others find that meditation works best right after sex. Explore your own responses and time your meditation practices accordingly.

Drugs are a more straightforward case. A drug is any substance, legal or not, that causes significant changes in the way your nervous system works. Here, the rule is simple: Drugs and meditation don't mix, and it's best to avoid meditation when you're under chemical influence. Again, that includes legal chemicals as well as illegal ones— coffee and alcohol as much as marijuana or cocaine.

The problems that arise from mixing drugs and meditation vary from drug to drug, but the basic principle is the same. Every drug disables some part of the complicated biochemical balance of your nervous system; that's how drugs work. At the very least, this makes meditation more difficult and less productive, and adding the effects of meditation to a nervous system in disarray is not a good idea. Some old meditation manuals state, for example, that meditating with alcohol in your bloodstream causes long-term neurological damage. While I don't know of any studies verifying this, it's not a risk you should take.

The basic rule is that you should avoid meditation until your system is free of drug effects, including any hangover. This can take twenty-four to forty-eight hours if you've used a lot of alcohol or any of the harder illegal drugs. There's much to be said, then, for the traditional rule that a student of meditation should avoid more potent drugs completely, and use milder ones such as coffee and wine very moderately. This may seem restrictive, but every path has its requirements, and one requirement of the Moon Path is a brain that's free of chemical fog. If the habit of using drugs is one you're not willing to break, or at least limit sharply, the Moon Path may not be for you.

POSTURE

Now that you've set aside space and time for meditation, and arranged other details of your life so they won't interfere with the practice, it's time to get to work. The first level of practice is purely physical: a posture of meditation that will allow you to remain comfortable, without moving, during each practice session.

Requirements for a meditation posture are straightforward. Your position should be stable and comfortable, so that you don't have to keep paying attention to your body. It should allow relaxation without encouraging sleep. It should also keep your spine vertical to foster deep breathing and to allow nwyfre to flow freely up and down the spinal channels.

Many people think of *padmasana*, the "lotus posture" of Hindu and Buddhist practice, whenever meditation postures come to mind. The lotus posture isn't appropriate for this work, however. It's uncomfortable for most Westerners and impossible for many; people have damaged their knees seriously trying to force themselves into it. Cross-legged postures such as padmasana also close off the flow of nwyfre between the meditator's body and the Earth, and this isn't helpful in Druid meditation—quite the opposite.

The two traditional positions for Druid meditation are a good deal easier to achieve. To enter the first and most common, sit down on a chair, far enough forward so that your back doesn't touch the back of the chair at all. The chair should face east when you're meditating, whenever that's possible, to take advantage of currents of nwyfre that circle the globe from east to west, driven by the play of forces between Earth and Sun. Facing into these currents while you meditate helps balance your subtle bodies.

Your posture should keep your thighs parallel to the ground and your hips and knees at right angles. If the seat of the chair is too low, put a pillow on it; if the seat is too high, put a book under your feet to raise your thighs to the proper angle. Your legs may be together or a short distance apart, as you prefer. Rest your hands palms down on your thighs, fingers pointing toward your knees. Keep your ankles directly under your knees, your feet flat on the floor (or the book), your elbows resting comfortably at your sides, your spine straight without being stiff, and the crown of your head level, as though you were balancing something on top of it. Keep your eyes open, looking ahead and slightly down.

Any posture that keeps the feet flat on the ground, the spine straight and uncompressed, and the breathing unrestricted can be used for meditation of the sort taught in this book. Some Druid traditions use standing postures of various kinds, and if standing meditation attracts you, fee free to try it. If you like, experiment with various positions. As you work your way through the beginning stages of the Moon Path though, it's best to choose one posture and stick with it, to make use of the power of habit.

The second meditation posture is less widely known. Many of the

Anglican meditation manuals used by Druids in the early years of the Revival gave special instructions for meditating while walking in a garden or some other quiet area. To meditate while walking, choose a route over level ground where you won't have to bend, climb stairs, duck around trees, or do anything else that will interrupt your thoughts. A paved or gravel path in a garden is ideal. It should lead in a circle, so that you can keep walking as long as necessary. Walk slowly and smoothly, taking relatively small steps at a steady rhythm. As with the seated posture, your spine should be straight without being stiff, the crown of your head level, and your eyes lowered. Let your arms move easily and naturally at your sides.

Spend your first practice session or two learning to settle into your chosen posture. If you're using the seated posture, sit down, arrange your body, and stay in the posture without moving for five minutes by the clock. This may turn out to be much more difficult than you expect. Even the most balanced posture will start to feel unfamiliar and awkward after a short time, and if the muscles around your spine are weak, your back may ache a little from the unfamiliar work of holding your body upright. You'll probably want to fidget and shift around, and you may start thinking about trying meditation in an over-stuffed easy chair instead.

Not so fast! Fidgeting is to the body what wandering thoughts are to the mind. Slumping back into an easy chair puts you in a position where you can't breathe deeply and won't benefit from the energy flows along the midline of your body. Stick with the posture, fight the urge to fidget, and don't let yourself wiggle, even when the posture seems intolerable. Most people find that, after some practice, discomfort fades and the posture becomes familiar and comfortable.

The rhythmic movement of the walking posture is often more comfortable, at least at first, but you should still spend a session or two practicing it. Over time, maintaining the slow steady pace needed for meditation can become quite as awkward as maintaining the motionless posture, and you may find yourself suddenly losing your balance and pitching to one side or the other as your body struggles with the posture. Keep with it and, in time, your body will find a new rhythm and move with the meditation, not against it.

RELAXATION

One of the things you'll notice when you try meditation is that your body is much more tense and uncomfortable than you ever imagined. Most people carry quite a lot of needless tension in their bodies. This excess tension is a problem for the meditator, since it distracts the mind and also gives rise to many physical and mental problems. The ability to notice and release tension is one of the positive side effects of meditation.

If you haven't done systematic relaxation before, set aside time outside your meditation sessions for relaxation practice. Start by getting into loose clothing and lying down on your back on a hard surface. This usually means the bare floor. A bed is too soft, and so is anything else padded. The surface should be hard enough to be uncomfortable—it's the discomfort that shows you where the tensions are. If the only hard surface available is bare concrete or stone, spread a thin blanket on it to keep it from absorbing heat too quickly from your body, since it's hard to relax when you're shivering.

You're likely to find this fairly unpleasant at first. The places that hurt are the places where you have plenty of excess tension built up. Do your best to put up with the discomfort; the first step in learning to relax is to pay attention to where you aren't relaxed.

Start by noticing how much of your body is actually touching the floor. Most people hold themselves up off the floor on a few points— the back of the head, the shoulder blades and upper back, the buttocks, the calves, the heels. If that's the way you're lying, imagine how much effort it's taking to support the weight of your body on those few points. That's a measure of the amount of energy you're putting into being tense.

Spend a few moments trying to sink down onto the floor, to broaden the areas of contact, so that more of you is resting on the floor. Don't try to do this by making an effort with your muscles! Instead, let your body spread out and sink; breathe slowly and evenly, and imagine tension leaving your body on each outbreath. Don't worry if you don't see much improvement at first. Like so much in the art of meditation, this has to be learned one step at a time.

Now turn your attention to your feet and relax them in the same way, breathing out tension on every outbreath. Go from your feet to your ankles, and do the same thing. From your ankles, go to your lower legs, your knees, your thighs, your buttocks, your groin, your hips, your abdomen (front and back together), your midriff (ditto), and your ribcage (ditto). Then start again at your hands: relax your hands, your wrists, your lower arms, your elbows, your upper arms, your shoulder joints, and the upper part of your torso where your shoulders join your chest. Then go on to your neck, your jaw, your face, and your scalp, finishing at the very top of your head.

The first few times you work your way through this process, you may notice only a small lessening of tension, or you may feel as limp as a boiled noodle. In either case, stretch, shift, and wiggle to wake up your muscles before you get up. You'll find that most of your tensions will slip right back into place before long, but not quite all of them will. By practicing this exercise regularly, you'll find that your body will gradually learn how to do without all that tension, and you'll see positive results in your meditation practice and the rest of your life.

In order to use the same skill in meditation, however, you must learn to let go of tension when you're in your meditation posture. This is a more complex matter, since you can't simply go limp without falling off the chair or onto the garden path. Your goal is to find a middle ground between tension and limpness; "poised" is a good word for the state you're trying to reach.

To learn this, once you've learned the meditation posture, put a few meditation sessions into learning how to relax in it. Sit down or start walking, and then go through the same process of step-by-step relaxation you did while lying on your back, starting with your feet and ending with your scalp. Take your time and release as much tension as you can.

As you do this, you may catch your thoughts drifting away from relaxation onto other subjects. Pay attention to this, and bring your attention back to the process of relaxation. Unless you have some experience with meditation, you'll probably spend a lot of time rounding up your thoughts and bringing them back to the job at hand. Don't get discouraged if this happens. This is actually a simple form of meditation. By focusing your awareness on tension and relaxation, you're already starting to meditate and building skills that will be central to all your Druid work.

BREATHING

Once you've learned posture and relaxation, the next step is breathing. What posture and relaxation do for your body, proper breathing does for nwyfre. This is crucial in meditation, since nwyfre, like your body, needs to be calm and still during meditative practices. An old metaphor describes it as a pool of water that can only reflect the surrounding trees and the sky when it is still. Stilling the pool of nwyfre is the task of the art of breathing.

The thought of breathing as an art may seem strange to you, and there's a reason for that. Most people in the modern world do a very poor job of breathing—precisely because we rarely give it any thought. We simply do it, taking the path of least resistance through whatever tensions and bad habits we've built up over the course of a lifetime. Getting out of the rut of habitual poor breathing and learning how to breathe fully and freely thus takes practice.

When you've learned a meditation posture and spent some practice sessions learning how to relax into it, devote your daily meditation sessions to breathing. Take up your meditation posture and relax into the position. It's not necessary to go through the whole feet-to-head relaxation drill, but relax your body overall. Spend a little time relaxing any areas that feel especially tense before you go on to breathing exercises.

There are three special breathing patterns used in this system of meditation. The first, called the cleansing breath, is done at the beginning and end of every meditation practice. Start by drawing in a slow, deep, full breath through your nose to the bottom of your lungs. As you breathe in, press out with your diaphragm muscles below your lungs; allow your belly and lower back to stretch outward, then your midriff and middle back, then your chest and upper back. At this point, your lungs should be full of air. Now purse your lips as though you were blowing through a straw, and breathe out in a series of short puffs. Keep puffing until your lungs are empty. Leave them empty for a moment or two, and then breathe in.

The cleansing breath, as the name suggests, cleans stale air and stagnant nwyfre from your lungs and body. It should not be overdone. Once at the beginning of each meditation session and once at the end is enough.

The second special breathing pattern, which follows the cleansing breath, is called the rhythmic breath. This is the workhorse of the system, so I will describe it in plenty of detail.

Start with your lungs empty, after a cleansing breath. Breathe in, slowly and smoothly, through your nose. Draw the breath down as far as possible at first, pressing down and out with your diaphragm muscles below your lungs, so that your belly and lower back expand while your chest remains motionless. Then expand the middle of your trunk, the area around your solar plexus, your floating ribs, and your middle back. Finally, once this is full, expand your chest and upper back, feeling the outward stretch of your ribs, your breastbone, and your shoulder blades.

At first, when you're learning the pattern, take these three phases one at a time. Later, with practice, they can become a single motion of the breath, which starts in your lower abdomen and rises up to your upper chest. Your lungs should be full, but not as full as in the cleansing breath. There should be room in your lungs for more air than you take in.

At the end of each inbreath, hold your breath for a short time by maintaining the expansion of the muscles of your trunk, not by closing your throat. If you hold your breath by closing your throat in breathing exercises, you risk damaging the sensitive tissues of your lungs. This point often takes practice, since many people's breathing muscles are weak from lack of use, and beginners sometimes draw in a deep breath only to feel it leaking away through their open throats while they try to hold on to it. Patience and practice will do the trick. If you have trouble telling whether your throat is open or closed, keep drawing a slow, thin current of breath in through your nose while you hold your breath.

Now breathe out through your nose in the same order you breathed in, pulling your belly and diaphragm inward and upward while keeping your chest expanded and drawing in your midriff. Then, finally draw in your chest. Do this slowly and smoothly, but not so slowly that you end up gasping.

Finally, when your lungs are completely empty, keep them that way for a short time. Again, do this by controlling the muscles of your chest and abdomen, not by closing your throat. If your breathing muscles are weak, keep breathing a thin stream of breath out through your nose during this phase. Begin again with an inbreath.

The rhythmic breath, as the name suggests, is done to a steady rhythm. The usual method is to breathe in while silently counting from one to three, hold the breath in while counting from one to three, breathe out while counting from one to three, and hold the breath out while counting from one to three. The counts should be evenly spaced and as slow as you can make them without puffing and gasping for air. If you're practicing walking meditation, the rhythm of your steps and your breathing should synchronize. With a little practice, you'll find a rhythm that works for you.

The third special breathing pattern is called the silent breath. This is done after the cleansing breath and the rhythmic breath during meditation. It's much simpler than the first two patterns, so you can concentrate on something other than breathing. The silent breath is done by breathing in slowly and smoothly in an ordinary rhythm. On the inbreath, draw the air down into the bottom of your lungs by gently pressing your diaphragm muscles down and out. On the outbreath, let your diaphragm rise up and in. Breathe through your nose slowly and gently, so that your breath makes no sound.

As your work with the silent breath deepens, you may find it useful to pay attention to the moment between the end of the inbreath and the beginning of the outbreath, and prolong it without actually holding your breath. In the Ancient Order of Druids in America, which borrowed the technique from modern Essene teachings, this moment is called the "Window of Opportunity" and is used to focus and recollect the self. Its deeper applications belong to more advanced levels of training than this book covers, but attention to the Window and its subtle effects will deepen your meditation and lay the groundwork for more advanced techniques further along your Druid path.

To practice breathing for meditation, assume your chosen posture and relax your body. Do the cleansing breath once and then start the rhythmic breath. Keep breathing with the rhythmic breath for at least five minutes. Focus your mind on breathing, bringing it back to your breath whenever it goes wandering off to some other subject. Then, when you're ready, shift to the silent breath and remain with it for as long as you wish, paying attention to the Window of Opportunity if that seems appropriate. Do one more cleansing breath to finish.

COLOR BREATHING

The rhythmic breath has another side, which provides one of the more subtle and flexible tools in the Druid toolkit. In Druidry, breath combines with imagination to work more intensely with nwyfre, in a technique called color breathing. Once you can do all three forms of breathing readily without having to look up the instructions, and have practiced them several times, you can learn color breathing.

Imagine that the space around you is filled with different colors of light. Like every Druid technique, this can be a challenge for some people at first, but anyone can learn to do it with a little practice. Take a moment to try it now, as you are reading. Imagine the space around you filling with clear green light, the color of sunlight through leaves. Imagine green all around you, like an ocean of color, reaching out infinitely in all directions. Pay particular attention to the space close to your body. Make sure the green color is all around you, coming right up to the boundary of your skin. Then, after you've built up the color as intensely as you can, dissolve it, until no trace of the green color remains in your awareness.

To practice color breathing, do this with a color and then imagine that you're breathing it into your body with each inbreath, and breathing it out with each outbreath, through your solar plexus—the area just below your breastbone, between the sides of your lower ribcage. As you breathe in, concentrate on the idea that the color is flowing into your body and filling it completely. In your mind's eye, see your whole body glowing intensely with the color. Then, as you breathe out, concentrate on the idea that every trace of the color is flowing out of you; see your body empty of the color, but the space around you filled with it.

The traditional explanation for this practice draws on old lore about the solar plexus. According to occult tradition, this is the place through which your body draws in aud, the solar current of nwyfre, from the Sun and the atmosphere. When you concentrate on a specific color in the air around you, the nwyfre surrounding your body takes on the energies associated with that color, and when you draw the charged nwyfre into yourself, the energies affect your mind and body. There are other, more psychological explanations. No matter how color breathing works, however, it *does* work, and can be a useful part of your meditation.

Your choice of color depends on the effect you want. The following colors and their uses are standard in Druid practice, but since everyone responds a little differently, you'll want to experiment with different colors yourself and note their effects.

>**Red** is energizing and awakening. Use when drowsiness is a problem in meditation, when practicing concentration (see pages 220–222), or when preparing for magical work. A clear rose red works well for this purpose. Use darker reds to energize your physical body.

>**Orange** combines the energizing effects of red with the clarity of yellow. Many people find a clear, pale orange best for ordinary meditation. Deeper oranges tend to encourage mental chatter and so should be avoided before meditation.

>**Yellow** clarifies and strengthens your mind. It's another common favorite for daily use in meditation; a light, clear yellow like morning sunlight is best. Deeper yellows, like deeper oranges, make the mind harder to direct.

>**Green** combines yellow's clarity with the receptive quali-

ties of blue. Green is commonly used before exercises that bring in nwyfre from the natural world (see page 230). Use pale spring greens before contacting solar energies and deep emerald greens for telluric contacts.

Blue is passive and receptive, opening the doors of awareness to a wider world. Blue color breathing helps prepare for divination, visionary work, and trance. The deeper the shade, the less conscious control of the experience you will have.

Violet combines active red and receptive blue into a color of pure power, potent but dangerous if mishandled. Violet is most often used for ritual magic and other intensive work with nwyfre, although it also has a place in techniques such as the Druid's Egg (see page 233).

White is like violet, but combines all the colors. Use it in working with gods and goddesses, and in other forms of theurgy or spiritual work. It should be clear and bright.

These are only general guidelines; there are many variations. The colors associated with the Druid elements and the Ogham letters can also be used in color breathing for particular effects. The eight stations of the year also have colors that relate, in turn, to energy centers within the subtle body. Study and practice will open these particular doors in due time.

When you meditate, do color breathing along with the rhythmic breath. After you assume your chosen posture, relax your body, and clear out stagnant nwyfre with the cleansing breath. Then imagine your chosen color all around you. Bring it in through your solar plexus as you breathe in, hold it in your body as you hold your breath, empty the color from your body as you breathe out, and be aware of it surrounding your body as you hold your lungs empty. When you finish the rhythmic breath, imagine the sea of color around you fading out of existence.

Concentration and Meditation

The next step in preparation moves from the realm of breath and nwyfre to the mind itself. Here, one of the main obstacles to meditation shows itself. Meditation requires the sustained movement of thought toward a chosen topic, but very few people can think about the same thing steadily for more than a few seconds at a time. The untrained mind—and nearly all of us have untrained minds—jumps from thought to thought like an overexcited monkey, never staying in one place for more than a moment. Thus, after you've learned to sit, to relax, and to breathe, you must learn to concentrate.

This is often the most frustrating part of learning to meditate; many people give up at this point. Dreams of magical power or spiritual wisdom clash uncomfortably with the realization that you can't keep your mind from misbehaving for five seconds in a row. Still, it's an obstacle that has to be faced. It helps to cultivate patience and the knowledge that everyone has to go through this stage.

Make a concentration target before beginning the exercise below. Find a piece of thin plywood or heavy cardboard six inches square. Paint it black, or glue a sheet of black paper onto it. In the center of the square, put a white circle one inch across; again, you can paint this or glue on a circle of white paper. The result, either way, is a plain white circle on a plain black background that, among other things, represents the Full Moon, the emblem of the lunar current in full manifestation. Place this in front of your meditation chair, below eye level, in a position where you will be looking directly at it during practice sessions.

Now you're ready to begin concentration training—a simple process, but not an easy one. Begin by assuming the sitting meditation posture, relax your body, take a cleansing breath to dispel stagnant nwyfre, and do about five minutes of the rhythmic breath with red color breathing. Then shift to the silent breath and turn your attention to the white circle in the center of the concentration target.

FIGURE 9. CONCENTRATION TARGET.

Your task is to keep your attention on that circle. Look at it, and be aware of yourself looking at it. Let your whole world contract to a white circle one inch in diameter. Unless you're unusually gifted, of course, this won't happen. What will happen instead is that you'll suddenly realize that you're thinking about something or looking somewhere entirely different, and you've completely lost track of the circle.

When this happens, don't just go back to the circle. Trace your thoughts back to where they left it. If you found yourself looking at your left knee, for instance, how did you get there? You were thinking about a bruise on that knee. Why? Because you got that bruise getting your cat down from a tree, and you were thinking about the cat. Why were you thinking about the cat? Because you remembered that you needed to go to the store to get cat food, and that memory, coming suddenly to mind, was what distracted you from the circle.

Your monkey-mind may have jumped through many more thoughts than this between the circle and wherever you caught up to it. Whatever the route, try to trace it all the way back to the point when your mind was distracted from the circle, and then resume focusing on the circle. When you catch your mind somewhere else a minute or two later, repeat the process, and repeat it, and repeat it, over and over, for at least ten minutes by the clock.

Your first few sessions of concentration are likely to be fairly demoralizing. For a while, you may even seem to be getting worse, not better, at keeping your attention on the circle. You may catch your mind jumping off in a new direction before you even get it back to the circle from an earlier distraction. In fact, entire sessions may go by with only a few seconds of attention to the circle—or none at all. Keep with it, however, and you'll notice three things happening. First, the time your attention stays fixed on the circle between monkey-expeditions will gradually increase. Second, the habit of returning to the circle along the path of your mind's movement will start to settle into place, so that your monkey-mind begins running back to the circle as enthusiastically as it runs away from it. Third, and most important, you'll find yourself starting to follow your mind as it moves, rather than losing track of the circle and not realizing it until later.

When your ten minutes of concentration includes at least five minutes of focused attention on the white circle, and when you can easily catch the movements of your mind and bring it back to the circle, you've learned what you need to learn from this exercise. Your preliminary training is finished, and you're ready to go on to meditation.

The difference between meditation and the exercises you've done so far is that meditation is directed toward an idea, a written sentence, an image, or something else you wish to understand. The subject of your meditation is called its theme. Anything at all can be a theme for meditation. Many spiritual schools have traditional sets of themes that students use as the basis for their meditations during training.

The material in Part II of this book is meant to provide you with themes for meditation. Each of the concepts from Druid symbolism and philosophy included there is intended to lead you to some of the subtler teachings and realizations of the Druid tradition. Don't be rigid about the choice of themes, however. If there are themes you particularly want to explore in your year of Druid training, by all means do so.

How do you select themes from a mass of written material like "Wisdom of the Stone Circle?" First, break larger themes that interest you up into pieces small enough for a single meditation. To start with, this usually means breaking themes into sentences and phrases, or selecting a series of sentences and phrases that contain the core ideas of the theme. One common mistake of novice meditators is to choose sprawling themes and hurry over them in a single session. More experienced meditators learn that one sentence or phrase is usually enough for one session. The point of the exercise is how much you can get from it, not how fast you can get through it.

Let's use the first of the nine Druid concepts from Part II as an example. You may be tempted to take the entire story of Einigen as the theme for a meditation, but that won't get you very far; it would be like trying to read and understand a long book by flipping through its pages in fifteen minutes. A more workable theme might be the first sentence of the tale:

> Einigen the Giant, the first of all beings, beheld three rays of
> light descending from the heavens at the beginning of the world.

That's a reasonable theme for a novice meditator. You may find, however, that even shorter themes give you more than enough material. This first sentence might more usefully be divided up into four themes:

> Einigen the Giant
>
> The first of all beings
>
> Three rays of light descending from the heavens
>
> The beginning of the world

Let's work with the first of these themes: Einigen the Giant. Select your theme no later than the night before you'll be meditating on it, because part of the preparation for meditating on a theme is, literally, to sleep on it. Just before you go to sleep, call it to mind, and let your mind rest on it for a while. Repeat it silently a few times to yourself.

Try to fall asleep with the words "Einigen the Giant, Einigen the Giant" murmuring in your mind. In fact, the simple act of introducing the words into your thoughts on the edge of sleep makes them part of the material your dreaming mind will mull over.

In the morning, at your usual time for meditation, begin meditating in exactly the same way you began the concentration exercises. Settle into the posture, relax, take one cleansing breath, and do about five minutes of rhythmic breath with appropriate color breathing. Then clear your mind completely for three silent breaths. If you find it useful, have your concentration target in front of you, just as in the concentration exercises, and keep your attention focused on it during those three final breaths. Your mind is already trained to focus on the white circle, and this can make it easier to keep your thoughts on track.

After the three silent breaths, turn your attention to the theme of the meditation. If the theme is expressed in words, repeat it three times slowly and silently in your mind. If it's an image, visualize it clearly in your mind's eye and hold the image for a few moments. In either case, concentrate on it as intently as you can, turning your attention away from everything else. The goal is to fix the theme firmly in your mind, so that you can return to it whenever you wish.

Now you begin the process of meditation. Start with the theme you've chosen, and think about it in a general way. Try to get a good overall sense of it. If it's a sentence or a phrase, go over it one word at a time, making sure you understand exactly what it says. If it's an image, pay careful attention to every detail of color and shape, every object in the picture.

A number of questions or issues are likely to come to mind as you consider your theme. You may wonder about the implications of a word in a sentence; you may puzzle over the meaning of a particular color or object in an image; you may consider the way some part of the theme relates to something in another meditation. This habit of wondering, of asking questions of yourself, is vital in meditation. Encourage it as much as you can.

Make a mental note of each question or issue you find in the theme. When you feel you have a general grasp of the theme, choose one of these questions and follow it out in thought as far as you can. Explore its meaning, its roots or sources, its consequences and implications. Turn it over and over in your mind, and try to figure out as much about it as you possibly can. Continue this for ten minutes, then take a cleansing breath to finish the meditation.

For example, in the case of our chosen theme, complete your pre-liminaries, then repeat the words "Einigen the Giant" three times—silently, slowly, and with total concentration. Now begin thinking about the theme. Unless you know quite a bit about Celtic linguistics, the name Einigen isn't likely to communicate much to you. The fact that Einigen was a giant is another matter. What is a giant? Why do giants play so large a role in legend? What roles do they play? How does Einigen relate to other giants of legend? What do giants symbol-ize? What does it mean that so many legendary beings are humanlike, but either much larger or much smaller than ordinary people? What does size mean in myth and legend generally?

These questions and others like them may come to mind as you con-sider the theme. Choose one of them: "How does Einigen relate to other giants of legend?" You may recall that most legendary giants are malevolent creatures fond of eating human beings. You may recall a book on Norse mythology and think of the giants of that tradition. They were all descended from the giant, Ymir, you remember, and Ymir was the first of all beings—just like Einigen. That may lead you to consider other mythologies you've encountered looking for other primordial giants. Though it's unlikely to happen in a single meditation, you may find yourself suddenly grasping why so many traditional teachings place a giant human form at the beginning of all things.

As you proceed with the meditation, there is one essential rule: *Be aware of exactly what you're thinking while you're thinking it*. This may seem like a strange rule, but it's the key to the entire method. Most people, most of the time, let thoughts go drifting unnoticed through their minds like leaves in an autumn wind—there one moment, gone the next, leaving no trace of themselves behind. (For example, what were you thinking about five minutes ago? Do you remember?) The heart of discursive meditation is the process of thinking consciously— and the only way to think consciously is to be conscious of what you're thinking.

As you try to do that, you may find yourself face to face with the same difficulties you encountered in the concentration exercises. Your thoughts may stray off in every direction you can imagine, and some you probably can't, instead of staying with the theme. Memories of the past, worries about the future, apparently random trains of thought set off by your surroundings or by nothing at all— these thoughts may all come crowding into your mind. You may realize that, for the last few minutes, you've been thinking of something completely unrelated to your theme. You may realize that, for the last few minutes, you haven't been thinking of anything at all.

The remedy is the same as in the concentration exercise. When your thoughts stray, bring them back to the point where you left your theme following the route along which they strayed. When you find that they've strayed off again, repeat the process, and repeat it, and repeat it.

On the other hand, you may find that you can't come up with any questions at all about your theme, or that the questions you do think of lead you nowhere. You may find yourself with nothing to think about at all, except the bare theme itself. This is another way that your mind tries to wiggle out of the hard work of meditation. When it happens, simply hold the theme itself in your mind. Do your best to keep your mind on it. If it's an image, try to see it clearly in your mind's eye. If it's a sentence, repeat it slowly in your mind. In either case, pay careful attention at each moment to what it means, rather than letting it turn into a meaningless pattern of shapes or sounds. Pursue the meaning, either until some new point comes up for you to think about, or until the meditation session is over.

If you're like most beginning students, you will probably waste seven of the ten minutes you devote to each practice session on stray thoughts unrelated to the topic or on blank periods when nothing comes to mind at all. You will probably spend another two minutes bringing your awareness back to your theme. That leaves a total of only one minute at most that will actually go to thinking about your theme. This is frustrating, but inevitable. Unless you've spent years in a monastery or a magical order, you're unlikely to know anything about how to use your own mind. That one minute of connected, meditative thinking—scattered and diffuse though it may seem—is the key to all the treasures of meditation. Scarcely less important are the two minutes you spend patiently bringing your thoughts back to heel. Keep working at it, walking the Moon Path, and the potentials of Druid meditation will unfold step by step before you.

Your second meditation should follow the same pattern, and the one after that, and so on, day after day. The material in Part II provides plenty of themes to start with. Once you've worked your way through those, let yourself be guided by your own inspiration or, as Druids often say, your own Awen. The books listed as further reading at the end of each part of this book can be rich sources of themes for meditation as well, as are the rituals in chapter 8. As you'll soon discover, you can find themes for meditation everywhere.

The things you learn about each theme constitute one of the benefits of meditation. Indeed, novices often look on them as meditation's most important benefit. Through meditation, the symbols and teachings of the Druid path will come to mean much more to you and will unfold many more of their secrets to you than they ever will to those who merely read about Druidry. Over time, however, the meaning of your themes will become less important than the experience of meditation itself. You will experience a slow but steady growth of inner power and wisdom that comes with regular practice. Let this happen in its own time; it can't be forced or faked, and the gradual unfolding is also relevant to the Druid path.

As your meditation practice deepens, you'll learn that improvement comes in cycles. Sometimes your meditations will be relatively successful, and other times not. Progress in meditation comes in fits and starts. At times, you may feel as if you've lost all the ground you had gained. It can be easy to slip into a mood of frustration or dejection when this happens, but the Druid teaching of the cycle of the year is relevant here: The bleak days before Imbolc, when it seems as though winter will last forever, come right before the budding of the first leaves of spring. Keep on meditating and the difficulties will resolve in time.

After a time, you may find that ten minutes of meditation no longer seems like enough. When you get to this point, add another five minutes, and stay at that level for a month or two at least before you add any more. If you still feel you need more time, add another five minutes, and stay at that level for another few months. This gradual, step-by-step expansion will help ensure that you don't leap beyond your actual abilities.

On the other hand, it's best never to cut back on the length of your practice sessions. If you've worked up to a twenty-minute session, that should be your absolute minimum from then on. To a surprising degree, progress in meditation comes from the simple fact of the time you spend doing it. Although the temptation to decrease the length of your practice sessions during difficult periods can be strong, the less time you spend meditating, the longer it will take you to work your way into a more productive phase.

Other Meditation Methods

Discursive meditation is the heart of the Moon Path, but other types of meditation have a place in Druid training as well. All of them involve working with awareness and nwyfre to heal and balance the self on many levels. Any such exercises known to the ancient Druids were lost centuries ago, so the Druids of the Revival did the sensible thing and borrowed effective methods from other traditions of spiritual development. Over the last three centuries, these borrowings have been reworked in the light of experience to form a toolkit of additional approaches. Here are some of the best methods for novice Druids.

RECEPTIVE MEDITATION

The discursive method of meditation described above differs sharply from most Asian meditative methods. Another Druid practice—receptive meditation—comes much closer to the disciplines of the East. Receptive meditation is best done outdoors in the presence of nature. The practice begins the same way discursive meditation does: Take your meditation posture, relax, and use color breathing—green in this case—to quiet your body and nwyfre. The method changes, however, when you get to the point of concentration.

In receptive meditation, when you reach the point of concentration, you must widen the focus of your mind instead of narrowing it. Try to be aware of everything around you, all at once. Open your awareness completely. Attend to sights, sounds, scents, air movements, and everything else, all together. If you've practiced the "splatter vision" technique covered back in chapter 7 use it here; let the entire universe enter into your awareness. If your mind starts to chatter—and of course it will in the early stages of your training—bring your attention back to the totality of the world around you. You're not trying to reach a conclusion or a realization, as in discursive meditation; you're simply trying to open yourself to the dance of the living Earth.

This form of meditation is a supplement to discursive meditation rather than a replacement for it. In many ways, it's simply an expansion of the Earth Path practice of stillness, and it makes a good supplement to your Earth Path work during and after your initiatory year. It can be practiced as often as circumstances permit. If you have the opportunity to plan a Druid retreat in a wilderness place, you can use long periods of receptive meditation to get the most out of the experience.

NWYFRE EXERCISES

The next three exercises bring nwyfre into your body from the outside. They can help you open yourself to deeper contact with living nature. All three can be strengthened by using color breathing immediately before starting. Most people find light green the best color to use.

Solar nwyfre exercise:

This exercise charges your body with solar nwyfre and attunes you with the solar current. Use it especially to dispell weakness, depression, and lack of energy. It's also good for building up health after an illness.

On a sunny morning between dawn and 10 A.M., find a place in the open air where you can stand in direct sunlight. Face the Sun, eyes closed, with as much of your body as possible in sunlight. Raise your arms to the sides and bend your elbows so that your arms form an arc, your fingertips pointing upward and your palms turned toward the Sun. Keep your arms as relaxed as possible and gently spread your fingers. Breathe slowly, feeling the sunlight soak into your body, and especially into your fingertips.

After one to three minutes, bring both hands inward to your head, stopping when they're a few inches from your forehead, with your palms facing you and your fingers nearly touching each other. Imagine the sunlight flowing into your head from your fingertips, illuminating the middle of your head.

Next, move your hands slowly down from your forehead to your solar plexus, maintaining the imagery of light pouring from your fingertips into the midline of your body. When your hands reach solar-plexus level, sweep them out to the sides and bow as low as you can, then return to the first position with your arms spread wide to greet the Sun. Repeat the entire process three or nine times.

Telluric nwyfre exercise:

This exercise is used to charge your body with telluric nwyfre and to attune with the telluric current. Use it especially to dispel tension, stress, and feelings of disconnection from nature, to release obsessive thoughts or memories, and to calm your mind after difficult times. The amount of time you practice it is less important than the place; it should be done standing on the bare earth, not on pavement or an artificial floor. It is best performed in a natural place where the land has suffered little or no damage at human hands. In fact, you'll find that the effects are different when you do the exercise on grass, bare soil, stone, or sand.

The exercise is essentially the solar nwyfre exercise in reverse. Hold your arms in a downward arc, hands about a foot out from your sides at hip level, your fingers pointing forward and your palms facing the ground. Again, relax your arms as much as possible and spread your fingers gently apart. Try to feel the presence of the Earth beneath you; with practice, you'll begin to perceive the telluric current itself flowing up from below. Breathe slowly and deeply for one to three minutes, feeling the Earth energies flowing into your fingertips.

Now bring your fingertips inward toward the vital center two inches below your navel, stopping a few inches from your body, palms toward you and fingertips nearly touching. Imagine the Earth currents flowing from your fingertips into the midline of your body. After a short time, move your hands slowly upward, maintaining the imagery of the telluric current pouring into the midline of your body from the vital center up to the throat center. At this point, separate your hands, sweeping them up and out to your sides, and bow deeply, then return to the first position with your hands turned toward the ground. Again, three or nine cycles of the exercise are enough.

Tree nwyfre exercise:

The third exercise works with the nwyfre of trees. Like Sun and Earth, trees of many species radiate nwyfre that benefits human beings. There's also another factor in play here: the trees benefit as well. Just as trees thrive on carbon dioxide, which human beings breathe out as a waste product, and produce oxygen, which human beings need to survive, so the excess nwyfre produced by many trees heals humans, while the nwyfre produced by humans nurtures trees. The same is true of other plants and many other things in the natural world. Indeed, an entire system of Druid natural medicine unfolds from the same principle.

Start the exercise by selecting a tree. Some varieties are considered more beneficial to human beings than others: pine, fir, and cedar are held to be best, with oak, beech, and apple not far behind. Elm is traditionally avoided, as its nwyfre is unhealthy for human beings.

Ask the tree's permission before exchanging nwyfre with it. Like other nonhuman creatures, trees don't understand human language, but they do understand our emotional patterns—often better than we do. Stand before the tree and ask it silently if it's willing to share its nwyfre with you. Feel for the answer and behave accordingly.

If the tree gives permission, approach it and come in contact with it. There are two traditional ways to do this. The first focuses the healing effects on your nervous system. To do this, stand or sit with your back pressed against the trunk. You can amplify the effect by putting your left hand on the small of your back with the palm turned out, touching the tree, and your right hand on your belly just below your navel, palm in.

The second method focuses the healing effects on the organs of your chest and abdomen. To do this, face the tree with your feet a few inches away from the base of the trunk, then lean the front of your body and your forehead against the tree. In this position, you can amplify the effect by placing your hands at forehead level, forming an upward-pointing triangle with your thumbs and index fingers, and spreading your other fingers out comfortably. Let your forehead rest against the tree in the center of the triangle. Bend your arms around the trunk, as though embracing the tree with your elbows.

In either posture, relax your body as much as possible and feel the tree as a living presence. An attitude of openness, friendship, and affection will yield the best results here. Ten or fifteen minutes of quiet presence next to a large, healthy tree will leave your own nwyfre cleansed and energized, and will do much the same for the tree.

THE DRUID'S EGG

The three exercises above draw nwyfre into your body. There are times, however, when it's necessary to do the opposite and seal out surrounding currents of nwyfre. Around angry people, for example, nwyfre takes on their emotions, and it's all too easy to be caught up in these negative feelings and drawn into an unwanted quarrel. The same thing can happen with other emotional states.

On a more extreme level, sites of suicide, murder, or other violent events can turn into dangerous pools of stagnant and disruptive nwyfre, where sensitive people can be overwhelmed. Such things play a much larger role in human life than most people ever suspect. You need serious magical training to handle the more extreme examples of this sort, but any Druid should know how to shield himself or herself from inappropriate nwyfre. The best way to do this is with an exercise called the Druid's Egg.

The Druid's Egg is best learned through meditation. Start by taking your meditation posture, relaxing your body, and taking a cleansing breath to clear stagnant nwyfre from your system. Next, spend about five minutes in violet color breathing before proceeding to three silent breaths, as though you were about to start a meditation.

Instead of calling a theme to mind, however, say silently to yourself: "I formulate the Druid's Egg." Imagine that the air suddenly becomes solid around you; see yourself at the center of a sphere of crystal—the Druid's Egg. Feel it around you, solid and firm, extending out several feet from your physical body. Pay attention to all sides of it—in front of you, behind you, to your left and right, above and below you. See it perfect and flawless, and concentrate on the idea that when the Druid's Egg is around you nothing can penetrate to harm you.

Hold the image in your mind as intensely as possible for several minutes, then say silently: "I release the Druid's Egg." As soon as you say this, imagine the crystal sphere dissolving back into thin air. Close with a single cleansing breath.

Practice this regularly, until the Druid's Egg formulates as soon as you silently repeat the words and stays without effort until you repeat the words that dissolve it. Once you can do this, you can call on the Egg outside of meditation, whenever it's necessary to screen out unwanted nwyfre. The Egg should only be kept in place for a short time, and should always be deliberately dissolved, so that you don't interfere with your body's ability to absorb and release nwyfre as needed.

AFTERWORD:

PATHS TO THE FOREST TEMPLE

THE YEARLONG INITIATION traced out in this book marks your first step into the world of Druidry. It doesn't complete the process of becoming a Druid—that lasts as long as you follow the Druid path. Nor is it anything more than a first brief introduction to learning about Druidry and the three paths that define its living core. It doesn't touch on the wealth of additional arts, crafts, traditions, and teachings that have gathered around Druidry since the dawn of the Revival. The image of the turning circle of time, which plays so important a part in Druid philosophy, has a lesson for us here. In your journeys in Druidry, just as in the Earth's journey through the stations of the year, every end is also a beginning. The end of your initiatory year is the beginning of a larger initiation—your journey into the wider realms of Druidry.

The route you take in that journey is yours to choose. While some spiritual traditions prescribe a fixed course of study from beginning to end with no room for individual choice or differences in personal needs, Druidry has long embraced a more open-ended approach. Most traditional Druid orders give novices a structured curriculum at first, but encourage more experienced students to take a progressively larger role in designing their own course of training, until finally, each one follows a unique path. This follows from the core principles of the Druid way. The point of Druid training, after all, isn't the mass production of initiates with identical skills and opinions, like so many cookies from a single cookie cutter. Rather, it's the awakening of the unique personal gifts and insights within each individual Druid.

For some Druids, this means that Druidry is a solitary path from beginning to end. There's nothing wrong with this choice. Becoming a Druid doesn't require membership in an organization, initiation rituals, or grandiose titles. It requires only reverence for the living Earth and the practice of a nature-centered spirituality and way of life. These can be done in perfect solitude.

For other Druids, however, taking part in the community of Druids is a welcome option. That community includes groups of all sizes and kinds. Some Druid groups are small, informal circles of friends who meet to celebrate the holy days and share explorations in the world of living nature. Others are international Druid orders with monthly newsletters, membership dues, training programs, and annual meetings at which hundreds of Druids gather from a dozen countries. Still others fall into the ample space between these two extremes. Depending on your needs and interests, any of these can be part of your journey in Druidry.

It's only fair to mention that the approach to Druidry taught in this book isn't welcome in all of today's Druid organizations. The legacy of the Druid Revival is still a hot button in some circles, and Druids who openly embrace the Revival's traditions come in for angry words now and then. Some groups on the Reconstructionist side of Druidry—those who attempt to reconstruct primitive Celtic religion as exactly as possible—still dismiss the entire Revival as "fake Druidry." Some nationalist Druids aligned with surviving Celtic cultures insist that nobody can be a Druid without Celtic ancestry or fluency in a Celtic language. Fortunately, most modern Druid organizations are more tolerant than this.

THE ANCIENT ORDER OF DRUIDS IN AMERICA

ONE ORGANIZATION THAT embraces the full heritage of the Druid Revival is the Ancient Order of Druids in America (AODA), which I presently head as Grand Archdruid. AODA began in 1912 as the American branch of the Ancient and Archaeological Order of Druids, a British Druid order founded in 1874 by the Rosicrucian mystic and Druid Robert Wentworth Little. In its early years, AODA had close links with Freemasonry, but it became a fully independent order in the course of the twentieth century. Today, it welcomes men and women of all religious, ethnic, cultural, national, and linguistic backgrounds, including Druids affiliated with other traditions and organizations within Druidry. AODA affirms Druidry as an inclusive path of nature spirituality, inspired by the ancient Druids and the traditions of the Druid Revival but oriented to the ecological and spiritual needs of the present and future. The book you're holding is a good general guide to its outlook and approach.

AODA offers three grades of initiation. Members join as Candidates and may remain at this level if they choose. Participation in the training program and advancement through the degrees is an option, not a requirement. Those Candidates who decide to seek advancement work through a year-long study program before receiving the first degree, the degree of Druid Apprentice. The study program includes the Earth, Sun, and Moon Paths and one of seven spirals—traditional arts long associated with Druidry. The book you're reading right now is the main First-Degree textbook.

Druid Apprentices who want to advance further work through a more advanced two-year curriculum to achieve the Second Degree, the degree of Druid Companion. Druid Companions who choose to go beyond this stage create and follow a three-year, self-designed study program to qualify for the Third Degree, the degree of Druid Adept. The curriculum for all three degrees may be found on the order's Web site at *http://www.aoda.org/about.htm*. The current First-Degree curriculum is included below.

Most members of AODA follow a solitary Druid path or work in small informal groups and initiate themselves through solo rituals provided by the Grand Grove, the governing body of the order. Second-degree members, however, have the right to organize local study groups that confer the Candidate and First-Degree initiations and practice other rituals. Third-Degree members have the right to organize local groves, which confer all the AODA degrees and perform every aspect of AODA ritual work. E-mail lists and other networking opportunities help AODA members at all levels work together as a community of like-minded Druids.

AODA asks a modest one-time fee for lifetime membership, and a similar fee for each of its three degrees of initiation. For details of these fees and the membership application process, as well as for other information about AODA, please visit our Web site at http://www.aoda.org or write to us at AODA, PO Box 1181, Ashland, OR 97520.

THE AODA
FIRST DEGREE
CURRICULUM

THE FOLLOWING STUDIES HAVE been established by the Grand Grove of AODA for the First Degree, the degree of Druid Apprentice. Under ordinary circumstances members who have enrolled in the Order and received the Candidate initiation must perform the following steps and wait a minimum of one year before receiving the First Degree. The Grand Grove may waive some or all of these requirements in special circumstances.

The First Degree curriculum consists of four parts. The Earth Path of nature awareness and service to the living Earth, the Sun Path of seasonal celebration, and the Moon Path of meditation are required of all Candidates. In addition, each Candidate must study one of the seven Spirals, studies which have traditionally been part of the Druid Revival—poetry, music, divination, healing, magic, sacred geometry, and Earth mysteries—and gain a basic level of competence in her or his chosen Spiral.

Candidates who wish to study more than one Spiral are encouraged to do so but this is not required for the First Degree.

Candidates should keep a Druid journal and note down details of all their Druid studies. The Druid journal need not be shown to anyone, including AODA officers, but functions as a resource for your First Degree examination. When you complete the First Degree study program, contact the Grand Grove for examination.

Updates and more information on the First Degree curriculum, including recommended reading lists for all Paths and Spirals, can be found online in the AODA Web site at *http://www.aoda.org*.

THE EARTH PATH

1. At least once each week during your Candidate year, spend fifteen minutes or more in direct contact with the natural world. This may be in a wild place (such as a forest or a seashore), in a place recovered by nature (such as an overgrown vacant lot), or in a place created by humanity and nature together (such as a garden or a park). Part of your time in nature should be spent practicing stillness, which simply involves sitting, keeping your mind empty of thoughts and distractions, and being wordlessly aware of everything around you. Part should be spent practicing focus, which involves detailed attention to some specific thing—a tide pool, a wild plant, the living things in a six-inch-square patch of grass, or the like.

2. Read at least nine books on nature, focusing on the natural history of the area in which you live, learning about the living things, the natural ecosystems and communities, the patterns of weather and water, and the development over time of the land. Relate as much of this information as possible to your own experiences of nature.

3. Make three changes in your lifestyle in order to take less from the Earth and give more back, and maintain those changes through your Candidate year. Different people lead different lives, and a change that would be easy for one might be difficult

or impossible for another; Druidry also affirms the need of individuals to make their own choices, so the choice of changes to make is left up to each Candidate. Whatever you choose should be something you're willing to keep doing for an entire year, and a small change you can sustain will do more good than a larger one that goes by the wayside when it proves unworkable.

4. Plant at least one tree during your Candidate year, and water and tend it until it is well established.

THE SUN PATH

1. During your Candidate year, celebrate a cycle of Druid holy days. In AODA the two solstices (around December 21 and June 21 each year) and the two equinoxes (around March 20 and September 23 each year) are traditional and should be part of your Druid calendar. Many members of the AODA also celebrate the "cross quarter days" of Imbolc (February 2), Belteinne (May 1), Lughnasadh (August 1), and Samhuinn (November 1), while others choose different days based on their own spiritual and cultural interests. The holy days may be celebrated alone or with others, using any ritual you prefer. Participating in community celebrations qualifies, as long as you're not simply a spectator. Write a detailed account of each celebration in your Druid journal, and write at least nine pages on the place of seasonal celebrations in your own Druid path and in the Druid tradition in general.

THE MOON PATH

1. Practice some form of meditation regularly during your Candidate year. While any form of meditation that involves focusing and directing the attention will qualify, discursive meditation is particularly recommended. Daily meditation using this or another approach is best, and while many people who are new to meditation need to work up to daily practice, this should be a goal to achieve by the end of your Candidate year.

2. The Sphere of Protection ritual should be learned and practiced daily during your Candidate year, or if you prefer another basic working to protect, balance, and purify your body of nwyfr, this may be substituted. Details of the Sphere of Protection can be obtained from the AODA Web site or by mail from the Grand Grove.

THE SEVEN SPIRALS

Spiral One: Poetry

1. Read and study poems by at least nine poets. Memorize three or more of these poems, totalling at least fifty lines.

2. Read at least three other books on the writing of poetry and use information from these in your own poetic practice. From these readings and your study of poetry, select at least thirty themes for meditation and use them in your daily meditations.

3. Keep a poet's journal. Set aside time at least once each week to write poems and do exercises.

4. Complete at least eighteen finished poems and revise them as needed. Nine of them must be in traditional poetic forms such as sonnet or haiku, while the others may be either formal or free verse as you prefer.

5. In your Druid journal, write at least three pages on the role you think poetry has in Druidry in general, and in your own Druid path.

Spiral Two: Music

1. Either select and acquire a musical instrument, or take up singing. Practice two hours a week as an absolute minimum—thirty to sixty minutes a day is a better target to aim for—using lessons, books, videos, and other sources of instruction to guide you. If you can attend at least one workshop on your chosen instrument during the year, do so.

2. Learn at least three solo pieces well enough that you would be willing to play them in public. If you have the chance, do so.

3. Read at least three books on music aside from your instructional books. From these readings, select at least thirty themes for meditation, and use them in your daily meditations.

4. Write at least three pages in your Druid journal on the place of music as a Druid art.

Spiral Three: Divination

1. Choose a method of divination and begin learning how to use it.

2. Perform daily divinations with your chosen method, using a simple spread, for at least nine months—in other words, at least 270 divinations. Keep a record of your daily divinations in your divination journal, and review them at intervals.

3. Perform at least nine other divinations, besides your daily readings, during your initiatory year.

4. Create a personal handbook of divination, writing out the meanings of each of the symbols in your divination method. Half a page per symbol is a good minimum.

5. Read at least three books on divination during your initiatory year. From this reading, select at least thirty themes for meditation, and use them in your daily meditations.

6. In your Druid journal, write at least three pages on the role you think divination has in Druidry in general, and in your own Druid path.

Spiral Four: Healing

1. Take classes to get certified in first aid and cardiopulmonary resuscitation.

2. Keep a health diary during your initiatory year. At least once each week, and more often whenever it seems useful, review your own current physical, emotional, and mental state. Keep track of anything that seems to affect your health in a positive or negative direction.

3. Select a healing art to study during your Candidate year. Choose and study at least three books on the subject, select from this reading at least thirty themes for meditation, and use them in your daily meditations.

4a. If your chosen healing art is herbalism, learn the properties and healing powers of nine common herbs. If at all possible, either grow them yourself or visit a place where they grow, and devote some of your Earth Path time to learning about them and their place in nature. Get to know them thoroughly and use them for your personal health care needs during your Candidate year. Make at least one infusion, one decoction, one tincture, and one salve. Any time you use one of the herbs for healing, take notes on your symptoms, how you treated them, and what happened.

4b. If your chosen healing art is something other than herbalism, do a comparable amount of work in that art; for example, if you decide to study acupressure, learn the routes of the meridians, memorize the location and healing effects of at least thirty important treatment points, use them as needed, and keep notes on your experiences.

4c. Whether you choose to study herbalism or some other healing art, a study program, correspondence course, or apprenticeship which includes at least an equivalent amount of work is a valid alternative to the program of independent study outlined here.

5. Write at least three pages in your Druid journal on the place of healing as a Druid art, in the tradition in general and in your own Druid path.

Spiral Five: Magic

1. Memorize a ritual for opening and closing magical space, and practice it at least once each week during your Candidate year. The AODA Grove opening and closing ritual is recommended for this purpose.

2. Using any preferred method, learn to summon, direct, and banish the energies of the three Druid elements until you can work with any of them at will.

3. Prepare and consecrate a wand, a cauldron, and a crane bag, as Druid magical working tools of nwyfre, gwyar, and calas respectively, using any preferred method.

4. Learn to place magical intentions in stones, as a way of making simple talismans and amulets. Do this at least three times for specific purposes, using any preferred method. Keep notes in your Druid journal on the results.

5. Read at least three books on magic during your initiatory year. From this reading, select at least thirty themes for meditation, and use them in your daily meditations.

6. In your Druid journal, write at least three pages on the place of magic in Druidry in general and in your personal Druid path.

Spiral Six: Sacred Geometry

1. Learn to construct a circle, a *vesica piscis*, an equilateral triangle, a square, a $\sqrt{2}$ rectangle, a $\sqrt{3}$ rectangle, and a Golden Rectangle on paper, using pen, straightedge and compass, without having to follow written instructions.

2. Do at least two of these constructions outdoors on the Earth, using rope and stakes instead of pen and ink. Describe the experiences in your Druid journal.

3. Read at least three books on sacred geometry. From these books, gather at least thirty themes for meditation, and use them in your daily meditations.

4. Set aside time to practice sacred geometry at least once each week during your Candidate year. Use books on sacred geometry as a source for constructions and diagrams to draw in your practice sessions.

5. In your Druid journal, write at least three pages on the role sacred geometry has in Druidry in general, and in your own Druid path.

Spiral Seven: Earth Mysteries

1. Acquire a set of dowsing tools, learn how to use them, and practice with them until you can get reliable results—for example, finding misplaced objects in your home.

2. Read at least three books on Earth mysteries. From these readings, select at least thirty themes for meditation and use them in your daily meditations.

3. At least once during your initiatory year, make a visit to a sacred site. Research the place in written sources, and learn as much as you can about it before you go there. See if it forms part of an alignment with other sites in the area. When you arrive, practice your Earth Path skills of stillness and focus. If at all possible, dowse the area for energy lines and currents within the Earth. Write about the experience in your Druid journal.

4. In your Druid journal, write at least three pages on the role you think Earth mysteries have in Druidry in general, and in your personal Druid path.

END NOTES

1. cited in Burl 1979, p. 42

2. Piggott 1975, p. 10

3. Piggott 1975, p. 181

4. Piggott 1975, p. 10.

5. Sallust, "On the Gods and the World" 1–4, in Sallust, *Sallust on the Gods* (Los Angeles. PRS, 1973).

6. cited in Ashe 1971, p. 14.

7. "The Marriage of Heaven and Hell," in *Complete Poems* (Penguin, 1977), p.194.

8. For more information on the Nature Challenge, including a detailed account of the science behind it, visit the David Suzuki Foundation website at http:www.davidsuzuki.org.

9. There are several different versions of this prayer, any of which can be used here. The most important variations are in the first and last lines. The oldest versions of the prayer, which date from the years when Christian Druidry was still the most common element in the Revival, begin "Grant, O God…" and end "…the love of God and all goodness." Later, less sectarian Druids said "Grant, O Divine Spirit…" or simply "Grant, O Spirit…" and used the same term in the last line. The revival of goddess spirituality in the last century saw some Druids start the prayer "Grant, O Goddess…" and end it "…the love of the Goddess and all goodness." Recent versions have been all over the theological map, and the form given above draws from one of these. Yet though different Druids invoke a remarkable assortment of spiritual powers, the things they ask for remain the same: protection and strength, understanding and knowledge, the ability to know what is right and to love it, and ultimately love for all existing things.

10. Roszak 1972, p. xvii.

BIBLIOGRAPHY

Adler, Margot. *Drawing Down the Moon*. New York: Penguin, 1997.

Ashe, Geoffrey. *Camelot and the Vision of Albion* London: Heinemann, 1971.

————. *Mythology of the British Isles*. North Pomfret, VT: Trafalgar Square, 1990.

Berger, Judith. *Herbal Rituals*. New York: St. Martins, 1998.

Blamires, Steve. *Celtic Tree Mysteries*. St. Paul, MN: Llewellyn, 1997.

Brown, Tom, Jr. *Tom Brown's Field Guide to City and Suburban Survival*. New York: Berkley, 1984.

————. *Tom Brown's Field Guide to Nature Observation and Tracking*. New York: Berkley, 1983.

————. *Tom Brown's Field Guide to The Forgotten Wilderness*. New York: Berkley, 1987.

Buhner, Stephen Harrod. *The Lost Language of Plants*. White River Junction, VT: Chelsea Green, 2002.

Calder, George. *Auraicept na n-Eces: The Scholars' Primer*. Edinburgh: John Grant, 1917.

Carr-Gomm, Philip. *The Druid Way*. Shaftesbury, Dorset: Element, 1993.

————. *The Elements of the Druid Tradition*. Shaftesbury, Dorset: Element, 1991.

————, ed. *The Druid Renaissance*. London: Thorsons, 1996.

Carr-Gomm, Philip and Stephanie Carr-Gomm. *The Druid Animal Oracle*. New York: Simon & Schuster, 1994.

————. *The Druidcraft Tarot*. London: Watkins, 2003.

Chadwick, Nora. *The Druids*. Cardiff: University of Wales Press, 1966.

Cook, Charles. *Awakening to Nature*. New York: Contemporary, 2001.

Cooper, Captain George H. *The Druid Bible*. San Jose, CA: Victor Hillis, 1936.

Critchlow, Keith. *Time Stands Still: New Light on Megalithic Science*. New York: St. Martins, 1982.

Daniel, Sir John. *The Philosophy of Ancient Britain*. Port Washington, NY: Kennikat Press, 1970.

Davies, Edward. *The Mythology and Rites of the British Druids*. London: J. Booth, 1809.

Elgin, Duane. *Voluntary Simplicity*. Rev. ed. New York: William Morrow, 1993.

Gardner, Adelaide. *Meditation: A Practical Study*. Wheaton, IL: Quest, 1968.

Geoffrey of Monmouth. *The History of the Kings of Britain*. London: Penguin, 1966.

Glass-Koentop, Pattalee. *Year of Moons, Season of Trees*. St. Paul: Llewellyn, 1991.

Graves, Robert. *The White Goddess*. New York: Farrar, Strauss and Giroux, 1966.

Graves, Tom. *Needles of Stone*. London: Turnstone, 1978.

Gray, William G. *Western Inner Workings*. York Beach, ME: Weiser, 1983.

Greer, John Michael *Earth Divination, Earth Magic*. St. Paul, MN: Llewellyn, 1999.

————. *Natural Magic*. St. Paul, MN: Llewellyn, 2000.

Grinsell, Leslie V. *Folklore of Prehistoric Sites in Britain*. London: David & Charles, 1976.

Guyonvarc'h, Christian-J. *The Making of a Druid: Hidden Teachings from The Colloquy of Two Sages*. Rochester, VT: Inner Traditions, 2002.

Hall, Manly Palmer. *Self-Unfoldment by Disciplines of Realization*. Los Angeles: Philosophical Research Society, 1942.

Hansen, Daniel. *American Druidism: A Guide to American Druid Groups*. Seattle, WA: Peanut Butter Publishing, 1995.

Hayman, Richard. *Riddles in Stone: Myth, Archaeology and the Ancient Britons*. London: Hambledon, 1997.

Heninger, S. K., Jr. *Touches of Sweet Harmony: Pythagorean Cosmology and Renaissance Poetics*. San Marino, CA: Huntington Library, 1974.

Hersey, George L. *Architecture and Geometry in the Age of the Baroque*. Chicago: University of Chicago Press, 2000.

Hitching, Francis. *Earth Magic*. New York: William Morrow, 1977.

Hopman, Ellen Evert. *A Druid's Herbal for the Sacred Earth Year*. Rochester, VT: Destiny, 1995.

————. *Tree Medicine, Tree Magic*. Custer, WA: Phoenix, 1991.

Hutton, Ronald. *Triumph of the Moon*. Oxford: Oxford University Press, 2000.

Jackson, Nigel and Nigel Pennick. *New Celtic Oracle*. Chieveley, Berks: Capall Bann, 1997.

Jenkins, Geraint H. *Facts, Fantasy, and Fiction: The Historical Vision of Iolo Morganwg*. Aberystwyth: Canolfan Uwchefrydiau Cymreig a Chgeltiadd Prifysgol Cymru, 1997.

Kaza, Stephanie. *The Attentive Heart: Conversations with Trees*. Boston: Shambhala, 1993.

Kelly, Fergus, trans. *Audacht Morainn*. Dublin: Dublin Institute for Advanced Studies, 1976.

Kendrick, T. D. *The Druids*. New York: Barnes and Noble, 1966.

Kindred, Glennie. *The Tree Ogham*. Sherwood, Nottinghamshire: By the author, 1997.

King, John. *The Celtic Druids' Year*. London: Blandford, 1994.

Kinkead, Eugene. *Wildness Is All Around Us*. New York: Dutton, 1978.

Knight, Gareth. *Occult Exercises and Practices*. Albuquerque, NM: Sun Chalice, 1998.

————. *The Practice of Ritual Magic*. Albuquerque, NM: Sun Chalice, 1998.

————. *The Secret Tradition in Arthurian Legend*. Wellingborough, Northants: Aquarian, 1983.

Leopold, Aldo. *A Sand County Almanac*. New York: Oxford University Press, 1949.

Lovelock, James. *Gaia: The Practical Science of Planetary Medicine*. Oxford: Oxford University Press, 2000.

The Mabinogion. trans. Jeffrey Gantz. London: Penguin, 1976.

MacCrossan, Tadhg. *The Sacred Cauldron*. St. Paul, MN: Llewellyn, 1991.

MacKie, Euan W. *Science and Society in Prehistoric Britain*. London: Paul Elek, 1977.

Malory, Sir Thomas. *Le Morte D'Arthur*. New York: Random House, 1994.

Matthews, Caitlin. *Celtic Wisdom Sticks: An Ogam Oracle*. London: Connections, 2001.

Matthews, John, ed. *The Bardic Source Book*. London: Blandford, 1998.

————, ed. *The Celtic Seers' Source Book*. London: Blandford, 1999.

————, ed. *The Druid Source Book*. London: Blandford, 1996.

McArthur, Margie. *Wisdom of the Elements*. Freedom, CA: Crossing, 1998.

McManus, Damian. *A Guide to Ogam*. Maynooth: An Sagart, 1991.

Merrifield, Ralph. *The Archeology of Ritual and Magic*. London: Batsford, 1987.

Michell, John. *A Little History of Astro-Archeology*. London: Thames & Hudson, 1989.

————. *City of Revelation*. New York: David McKay, 1972.

————. *The Dimensions of Paradise*. London: Thames & Hudson, 1988.

————. *The Earth Spirit: Its Ways, Shrines and Mysteries*. New York: Crossroad, 1975.

————. *Megalithomania*. London: Thames & Hudson, 1982.

————. *The New View Over Atlantis*. London: Thames & Hudson, 1983.

————. *The View Over Atlantis*. New York: Ballantine, 1969.

Miles, Dillwyn. *The Secret of the Bards of the Isle of Britain*. Llandebie, Wales: Gwasg Dinefwr, 1992.

Mountfort, Paul Rhys. *Ogam: The Celtic Oracle of the Trees*. Rochester, VT: Destiny, 2002.

Murray, Colin. *Golden Section Order Broadsheets*. London: Cantata Organica Press, 1956–1977.

Murray, Liz and Colin. *The Celtic Tree Oracle*. New York: St. Martin's, 1988.

Naddair, Kaledon. *Ogham, Koelbren, and Runic*. Edinburgh: Keltia Publications, 1987.

Nicholas, T. Islwyn. *A Welsh Heretic: Dr. William Price, Llantrisant*. London: Foyle's, 1941.

Nichols, Ross. *The Book of Druidry*. London: Thorsons, 1990.

Nollman, Jim. *Spiritual Ecology*. New York: Bantam, 1990.

North, John. *Stonehenge: A New Interpretation of Prehistoric Man and the Cosmos*. New York: Free Press, 1996.

O'Boyle, Sean. *Ogam: The Poet's Secret*. Dublin: Gilbert Dalton, 1980.

O'Flaherty, Roderic. *Ogygia*. Dublin: W. M'Kenzie, 1793.

Owen, A. L. *The Famous Druids*. Oxford: Clarendon Press, 1962.

Paterson, Helena. *The Celtic Lunar Zodiac*. St. Paul, MN: Llewellyn, 1998.

————. *The Handbook of Celtic Astrology*. St. Paul, MN: Llewellyn, 1999.

Pennick, Nigel. *The Ancient Science of Geomancy*. London: Thames and Hudson, 1979.

————. *Earth Harmony*. London: Century, 1987.

————. *Games of the Gods*. York Beach, ME: Weiser, 1989.

Pettis, Chuck. *Secrets of Sacred Space*. St. Paul: Llewellyn, 1999.

Piggott, Stuart. *The Druids*. London: Thames & Hudson, 1975.

————. *William Stukeley*. Rev. ed., London: Thames & Hudson, 1985.

Pitts, Mike. *Hengeworld*. London: Century, 2000.

Plotnik, Arthur. *The Urban Tree Book*. New York: Three Rivers, 2000.

Poynder, Michael. *Pi in the Sky: A Revelation of the Ancient Celtic Wisdom Tradition*. Cork: Collins, 1997.

Pryor, Francis. *Seahenge: New Discoveries in Prehistoric Britain*. London: HarperCollins, 2001.

Raftery, Barry. *Pagan Celtic Ireland: The Enigma of the Irish Iron Age*. London: Thames and Hudson, 1994.

Rees, Alwyn and Brinley Rees. *Celtic Heritage: Ancient Tradition in Ireland and Wales*. London: Thames and Hudson, 1961.

Roberts, Anthony. *Sowers of Thunder: Giants in Myth and History*. London: Rider, 1978.

Roszak, Theodore. *Where the Wasteland Ends*. New York: Anchor, 1972.

Roy, Rob. *Stone Circles: A Modern Builder's Guide to the Megalithic Revival*. White River Junction, VT: Chelsea Green, 1999.

Mouni Sadhu. *Concentration*. No. Hollywood, CA: Wilshire, 1959.

————. *Meditation: An Outline for Practical Study*. No. Hollywood, CA: Wilshire, 1967.

Skeels, Dell. *The Romance of Perceval in Prose*. Seattle: University of Washington Press, 1961.

Skelton, Robin. *Spellcraft*. Toronto: McClelland and Stewart, 1978.

Spence, Lewis. *The History and Origins of Druidism*. London: Rider & Co., 1949.

————. *The Magic Arts in Celtic Britain*. Reprint. Minneola, NY: Dover, 1999.

————. *The Minor Traditions of British Mythology*. Reprint. New York: Arno Press, 1979.

Steiner, Rudolf. *The Druids*. Forest Row, E. Sussex: Sophia, 2001.

————. *How to Know Higher Worlds*. Hudson, NY: Anthroposophic Press, 1994.

————. *A Way of Self-Knowledge*. Hudson, NY: Anthroposophic Press, 1999.

R. J. Stewart. *The Mystic Life of Merlin*. New York: Arkana, 1986.

————. *The Prophetic Vision of Merlin* New York: Arkana, 1986.

————. *The Spiritual Dimension of Music*. Rochester, VT: Destiny, 1987.

————. *The UnderWorld Initiation*. Wellingborough, Northamptonshire: Aquarian, 1985.

————. *The Way of Merlin*. London: Aquarian, 1991.

Stirling, William. *The Canon*. Reprint. York Beach, ME: Weiser, 1999.

Strachan, Gordon. *Jesus the Master Builder: Druid Mysteries and the Dawn of Christianity*. Edinburgh: Floris, 1998.

Street, Christopher. *Earth Stars, the Visionary Landscape, Part One: London, City of Revelation*. London: Hermitage, 2001.

Temple, Robert. *The Crystal Sun*. London: Century, 2001.

Thompson, D'Arcy W. *On Growth and Form*. Cambridge: Cambridge University Press, 1971.

Thorsson, Edred. *The Book of Ogham*. St. Paul, MN: Llewellyn, 1994.

Nikolai Tolstoy. *The Quest for Merlin*. Boston: Little, Brown, and Co., 1985.

Underwood, Guy. *The Pattern of the Past*. New York: Abelard-Schumann, 1973.

Watkins, Alfred. *The Old Straight Track*. London: Methuen, 1925.

Wheeler, James. *The Modern Druid*. London: n.p., 1743.

Williams, Taliesin. *The Iolo Manuscripts*. Llandovery: William Rees, 1848.

Williams ab Ithel, Rev. J., ed. and trans. *Barddas*. Llandovery: D.J. Roderic, 1862.

Wood, Ernest. *Concentration: An Approach to Meditation*. Wheaton, IL: Quest, 1949.

Wright, Dudley. *Druidism, the Ancient Faith of Britain*. London: E.J. Burrow, 1924.

INDEX

TO OUR READERS

Weiser Books, an imprint of Red Wheel/Weiser, publishes books across the entire spectrum of occult and esoteric subjects. Our mission is to publish quality books that will make a difference in people's lives without advocating any one particular path or field of study. We value the integrity, originality, and depth of knowledge of our authors.

Our readers are our most important resource, and we appreciate your input, suggestions, and ideas about what you would like to see published. Please feel free to contact us, to request our latest book catalog, or to be added to our mailing list.

Red Wheel/Weiser, LLC
500 Third Street, Suite 230
San Francisco, CA 94107
www.redwheelweiser.com